BIRKHÄUSER

Frontiers in Mathematics

Vladislav V. Kravchenko

Applied
Pseudoanalytic
Function
Theory

Birkhäuser Verlag
Basel · Boston · Berlin

Author:
Vladislav V. Kravchenko
Departamento de Matemáticas
CINVESTAV del IPN
Unidad Querétaro
Libramiento Norponiente # 2000
Fraccionamiento Real de Juriquilla
Santiago de Querétaro, Qro., C.P. 76230
México
e-mail: vkravchenko@qro.cinvestav.mx

2000 Mathematical Subject Classification: 30G20, 30G35, 34B05, 34B24, 35B05, 35C05, 35C10, 35J05, 35J10, 35J15, 35J25, 35L10, 35Q35, 35Q40, 35Q60, 78A30, 78M25, 81Q05, 81Q60

Library of Congress Control Number: 2009925158

Bibliographic information published by Die Deutsche Bibliothek
Die Deutsche Bibliothek lists this publication in the Deutsche Nationalbibliografie; detailed bibliographic data is available in the Internet at <http://dnb.ddb.de>.

ISBN 978-3-0346-0003-3 Birkhäuser Verlag AG, Basel · Boston · Berlin

© 2009 Birkhäuser Verlag AG
Basel · Boston · Berlin
P.O. Box 133, CH-4010 Basel, Switzerland
Part of Springer Science+Business Media
Cover design: Birgit Blohmann, Zürich, Switzerland
Printed on acid-free paper produced from chlorine-free pulp. TCF ∞

ISBN 978-3-0346-0003-3   e-ISBN 978-3-0346-0004-0

9 8 7 6 5 4 3 2 1   www.birkhauser.ch

To Kira

Contents

Part I Pseudoanalytic Function Theory and Second-order Elliptic Equations

Part III Applications to Real First-order Systems

9 Beltrami Fields

10 Static Maxwell System in Axially Symmetric Inhomogeneous Media

Part IV Hyperbolic Pseudoanalytic Functions

11 Hyperbolic Numbers and Analytic Functions

12 Hyperbolic Pseudoanalytic Functions

13 Relationship between Hyperbolic Pseudoanalytic Functions and Solutions of the Klein-Gordon Equation

Part V   Bicomplex and Biquaternionic Pseudoanalytic Functions and Applications

Foreword

This book is about applications of the theory of pseudoanalytic functions. The Swedish mathematician T. Carleman was the first to characterize this class of functions via a generalized Cauchy-Riemann system which is sometimes called a Carleman-Vekua-Bers system, as the investigation of the solutions of these systems reached fruition many years later through work of I.N. Vekua and L. Bers. In the 1940s, L. Bers accepted an invitation to participate in the program Advanced Research and Instruction in Applied Mathematics at Brown University. There, as part of work relevant to the war effort, he studied two-dimensional subsonic fluid flow problems, which led him to pseudoanalytic functions. Independently, I.N. Vekua from Tiflis called them "generalized analytic functions" and described applications in elasticity and fluid dynamics. Under the influence of his famous book, *Generalized Analytic Functions* (1959), operator theoretical aspects dominated for a time, while Bers' ideas of a theory similar to classical complex analysis fell into relative oblivion.

The current renaissance of Bers' theory is primarily due to recent research by V.V. Kravchenko, who independently and later in collaboration with other colleagues uncovered striking new relations and applications of pseudoanalytic function theory. These developments have now been very carefully prepared and presented in a style accessible to a wide audience. Through several interesting examples from physics it is shown how concepts of Bers' theory give new results. The book is an interplay between pseudoanalytic theory and a collection of partial differential equations of mathematical physics which are quite important in applications: inter alia, the Schrödinger equation, the Klein-Gordon equation, Maxwell's equations and the Dirac equation. One basic idea which can be found throughout the book is that, starting from a special solution, it is often possible to construct large classes of solutions and even complete systems of solutions to an important equation. Mathematical topics are motivated by physical problems, each leading to a corresponding Carleman-Vekua-Bers system that is the main subject of this book.

The reader will find in this book suprising relations among equations from different genres. To fully appreciate them, some knowledge is required of complex analysis, ordinary differential equations including Sturm-Liouville problems, and second-order elliptic partial differential equations. Researchers may take note that the theory presented in this book is not yet complete; new applications can easily be visualized, and the author has formulated a number of open problems which may be tackled in the future. A comprehensive list of papers and books is provided at the end of the volume, suited for readers who wish to deepen their research studies. Kravchenko's book is to be recommended to higher undergraduates, graduates and postdoctoral researchers.

Wolfgang Sproessig

Freiberg, December 2008

Chapter 1

Introduction

Pseudoanalytic function theory generalizes and preserves many crucial features of complex analytic function theory. The Cauchy-Riemann system is replaced by a much more general first-order system with variable coefficients. The foundations of pseudoanalytic function theory have been created by a considerable number of mathematicians among whom Lipman Bers and Ilya Vekua played the most prominent role. In the book of I. Vekua [120] and by many other researchers, pseudoanalytic functions are called generalized analytic. Nevertheless in the present work we implement the term "pseudoanalytic" in order to emphasize the fact that we mainly use the part of the theory developed by L. Bers and his collaborators [13].

Pseudoanalytic function theory found many applications in different fields of mathematics and mathematical physics. Historically it became one of the important impulses for developing the general theory of elliptic systems. Here, the Vekua theory played a more important role due to its tendency to a more general, operational approach. L. Bers tried to follow more closely the ideas of classical complex analysis and paid more attention to the efficient construction of solutions. Among other results, L. Bers obtained analogues of the Taylor series for pseudoanalytic functions and some recursion formulae for constructing generalizations of the base system $1, z, z^2, \ldots$. The formulae require knowledge of the Bers generating pair (two special solutions) of the corresponding Vekua equation describing pseudoanalytic functions as well as generating pairs for an infinite sequence of Vekua equations related to the original one. The necessity to count with an infinite number of exact solutions of different Vekua equations turned out to be an important obstacle for efficient construction of Taylor series (in formal powers) for pseudoanalytic functions.

Traditional applications of pseudoanalytic functions include boundary value problems in elasticity theory and hydrodynamics [120]. This book is dedicated to other applications. In the recent works of the author [68], [69] and [71] a close connection between the second-order elliptic equation

$$(\operatorname{div} p \operatorname{grad} + q)u = 0 \tag{1.1}$$

and a Vekua equation of a special form was presented. This connection is a direct generalization of a relation which exists between harmonic and analytic functions. The special form of the arising Vekua equation (which we call the main Vekua equation) allows us to apply well-developed methods of pseudoanalytic function theory ([6], [13], [17], [32], [116], [120], [123] and others) and of p-analytic function theory [103] to the analysis of the corresponding second-order equations.

In the first part of this book, in Chapters 3 and 4 we develop the theory of series in formal powers for the main Vekua equation and as a consequence the corresponding theory for equation (1.1). Formal powers were defined by L. Bers (see [13]) and represent a generalization of the usual powers $(z - z_0)^n$ which play a crucial role in the one-dimensional complex analysis. As their name reveals, formal powers in general are not powers. They behave like $(z - z_0)^n$ only locally, near the center, and in fact can be complex functions of a quite arbitrary nature. Nevertheless they are solutions of a corresponding Vekua equation and under quite general conditions represent a complete system of its solutions in the same sense as any analytic function under quite general and well-known conditions can be approximated arbitrarily closely by a normally convergent series of complex polynomials.

Extending the original results of L. Bers we present a simple procedure for explicit construction of formal powers corresponding to the main Vekua equation in a very general situation [73]. From the relation of the main Vekua equation to equation (1.1) we obtain that, under quite general conditions, we are able to construct explicitly a complete system of exact solutions of (1.1). More precisely, let us consider, e.g., the conductivity equation

$$\operatorname{div}(p \operatorname{grad} u) = 0.$$

Our result then gives us the possibility to construct explicitly a complete system of solutions of this equation if p has the form

$$p = \Phi(\varphi)\Psi(\psi) \tag{1.2}$$

where (φ, ψ) is an orthogonal coordinate system, Φ and Ψ are arbitrary positive differentiable functions.

In the case of the stationary Schrödinger equation

$$(-\Delta + q)u = 0, \tag{1.3}$$

in order to construct a complete system of solutions explicitly we need a particular solution of this equation of the form (1.2). Note that before this result had been obtained the knowledge of one particular solution of a second-order equation in two dimensions like (1.3) had not given much information about the general solution. We show that one particular solution of (1.3) generates a complete system of solutions of (1.3) which in a sense and for many purposes represents the general solution of the equation. We give an introduction to this new method and include explanation of the theory behind it and some examples of application including a numerical method for solving boundary value problems for (1.1) based on the construction of pseudoanalytic formal powers.

The arsenal of pseudoanalytic function theory includes the integral representations for pseudoanalytic functions among which the Cauchy integral formula is of great importance. As in the case of explicit construction of formal powers, considerable difficulties arise in the explicit construction of the Cauchy integrals. Only in some very special cases was the pseudoanalytic Cauchy integral obtained explicitly. Recently, in [72] a certain progress in this direction was achieved. A procedure for explicit construction of the Cauchy kernels for an important class of pseudoanalytic functions was developed. We present it in Chapter 5.

Behind these recent developments in pseudoanalytic function theory there is a deep and quite universal idea of factorization. The special Vekua equations closely related to second-order elliptic equations arise as factorizing terms in a factorization of the operator in (1.1). In relation to this factorization a nonlinear complex equation of a special form appears which, as we show in Chapter 6, enjoys many important properties of the ordinary differential Riccati equation. We prove the generalizations of the famous Euler and Picard theorems as well as some new features arising in the complex situation.

Another new application of pseudoanalytic function theory is considered in the second part of the present book and corresponds to the theory of linear second-order ordinary differential equations. The problem of solving the Sturm-Liouville equation

$$(pu')' + qu = \lambda r u \qquad (1.4)$$

by a known nontrivial solution of the equation

$$(pu_0')' + qu_0 = 0 \qquad (1.5)$$

where p, q, r, u, u_0 are complex-valued functions of the real variable x and λ is an arbitrary complex constant is of fundamental importance due to numerous situations in mathematical physics where it arises. For example, when the method of separation of variables is applied to the equation

$$\operatorname{div}(P\nabla v) + Qv = 0$$

where P and Q possess some symmetry sufficient for separating variables, very often one can arrive at the equation (1.4), and it is really desirable to have a possibility to solve only one equation (1.5) and to derive from its solution the solution of (1.4). Moreover, in many important cases the solution of (1.5) is known. For example, consider the conductivity equation

$$\operatorname{div}(P\nabla v) = 0$$

and suppose, e.g., that P is a function of one Cartesian variable (for a recent work motivating this example see [35]). Separation of variables leads to the equation

$$(Pu')' = \lambda u$$

and the solution of the corresponding equation (1.5) is given, a particular nontrivial solution can be chosen as $u_0 \equiv 1$. Thus, to have a method allowing us to transform u_0 into u means a complete solution of the original problem.

There are dozens of works dedicated to the construction of zero-energy solutions of the Schrödinger equation (see, e.g., [23], [24]). With the aid of the results of the present work these solutions can be used for obtaining solutions for all other values of λ. Moreover, the presented result is directly applicable to Dirac systems with scalar potentials as was observed in [62]. These are just some immediate applications of this result.

We obtain a new representation for solutions of Sturm-Liouville equations and consider its applications to the solution of initial-value and spectral problems. This new representation lends itself to numerical computation representing a new and efficient numerical method for solving Sturm-Liouville problems

The methods of pseudoanalytic function theory presented in this book are applicable also to important systems of mathematical physics. In the third part we show how the system describing so-called Beltrami fields as well as the static Maxwell system for inhomogeneous media can be treated using this technique. The result is that under quite general conditions a complete system of solutions of both systems can be obtained explicitly and used for solving corresponding boundary value problems.

If instead of complex numbers the algebra of hyperbolic numbers is used, a good deal of pseudoanalytic function theory can be developed along the same lines. In this case we obtain hyperbolic pseudoanalytic function theory which we present in Part IV. Instead of the factorization of elliptic second-order operators we obtain the factorization of the hyperbolic operators, in particular of the Klein-Gordon operator. As in the elliptic case, formal powers and corresponding solutions of the Klein-Gordon equation can be obtained explicitly. We present this result in Chapter 13.

In Part V we discuss other generalizations of pseudoanalytic function theory. First of all, the development of a bicomplex generalization of this theory due to applications to second-order equations with complex coefficients as well as to such objects as the Dirac equation turns out to be very important. Second, it is clear that some important facts from pseudoanalytic function theory can be generalized to a multidimensional situation using quaternions or more general Clifford algebras. We introduce the bicomplex Vekua equation and basic concepts of bicomplex pseudoanalytic function theory. We show how the Dirac equation with electromagnetic and scalar potentials is related to this theory. In a special case of a scalar potential we obtain an infinite system of exact solutions, once more using the theory of pseudoanalytic formal powers.

In Chapter 16 we consider second-order elliptic equations in a three-dimensional case using quaternionic-analytic tools. The quaternionic factorization of the Schrödinger operator leads us to a spatial generalization of the main Vekua equation which possesses properties similar to those of the complex Vekua equation.

Obviously, the results presented in this book are far from being complete or final. Much can be done in different directions. This is why we included a section with a discussion of some open problems related to the results presented in this book.

The author expresses his gratitude to CONACYT (Mexico) for partial support of this work.

Querétaro, December 2008

Part I

Pseudoanalytic Function Theory and Second-order Elliptic Equations

Chapter 2

Definitions and Results from Bers' Theory

This chapter is based on notions and results presented in [13] and [14]. Let Ω be a simply connected domain in $\mathbf{R}^2$. We use the complex variables $z = x + iy$, $\zeta = \xi + i\eta$, $w = u + iv$, etc. Complex conjugates are denoted by bars $\overline{z} = x - iy$. Sometimes we will use the operator of complex conjugation: $Cw = \overline{w}$. Functions of x and y are written as functions of z without implying analyticity. We denote $\partial_{\overline{z}} = \frac{1}{2}\left(\frac{\partial}{\partial x} + i\frac{\partial}{\partial y}\right)$ and $\partial_z = \frac{1}{2}\left(\frac{\partial}{\partial x} - i\frac{\partial}{\partial y}\right)$. The notation $w_{\overline{z}} = \partial_{\overline{z}}w$ and $w_z = \partial_z w$ will also be used throughout the book.

2.1 Generating pairs and differentiation

The starting point of Lipman Bers' theory of pseudoanalytic functions is the notion of a generating pair which is a couple of complex functions, independent in the sense that at any point the value of any complex function defined there can be represented as a real linear combination of the generating functions. In pseudoanalytic function theory they play the same role as 1 and i in the theory of analytic functions.

Definition 1. A pair of complex functions F and G in Ω, possessing Hölder continuous[1] partial derivatives with respect to the real variables x and y, is said to be a generating pair if it satisfies the inequality

$$\operatorname{Im}(\overline{F}G) > 0 \qquad \text{in } \Omega. \tag{2.1}$$

It follows that for every z_0 in Ω and any complex function w defined in z_0 we can find unique real constants λ_0 and μ_0 such that $w(z_0) = \lambda_0 F(z_0) + \mu_0 G(z_0)$.

---

[1]See the definition of Hölder continuity in Section 4.3.

Definition 2. Let the function w be defined in a neighborhood of z_0. We say that at z_0 the function w possesses the (F, G)-derivative $\dot{w}(z_0)$ if the (finite) limit

$$\dot{w}(z_0) = \lim_{z \to z_0} \frac{w(z) - \lambda_0 F(z) - \mu_0 G(z)}{z - z_0} \tag{2.2}$$

exists.

Sometimes instead of the notation $\dot{w}$ for the (F, G)-derivative of w we will use the notation $\frac{d_{(F,G)}w}{dz}$.

Set (for a fixed point z_0)

$$W(z) = w(z) - \lambda_0 F(z) - \mu_0 G(z),$$

the real constants λ_0 and μ_0 being uniquely determined by the condition

$$W(z_0) = 0.$$

The function W has partial derivatives if and only if w has, and $\dot{w}(z_0)$ exists if and only if $W'(z_0)$ does. Moreover, if it exists then $\dot{w}(z_0) = W'(z_0)$ where $W'(z_0)$ is the complex derivative of W at the point z_0: $W'(z_0) = \lim_{z \to z_0} \frac{W(z) - W(z_0)}{z - z_0}$. Hence the existence of $W_z(z_0)$, $W_{\bar{z}}(z_0)$ and the equation

$$W_{\bar{z}}(z_0) = 0 \tag{2.3}$$

are necessary for the existence of $\dot{w}(z_0)$, and the existence and continuity of $W_z(z)$, $W_{\bar{z}}(z)$ for $|z - z_0| < r$ together with (2.3) are sufficient.

The function W can be represented in the form

$$W(z) = \frac{\begin{vmatrix} w(z) & w(z_0) & \overline{w(z_0)} \\ F(z) & F(z_0) & \overline{F(z_0)} \\ G(z) & G(z_0) & \overline{G(z_0)} \end{vmatrix}}{\begin{vmatrix} F(z_0) & \overline{F(z_0)} \\ G(z_0) & \overline{G(z_0)} \end{vmatrix}}, \tag{2.4}$$

so that (2.3) can be written as

$$\begin{vmatrix} w_{\bar{z}}(z_0) & w(z_0) & \overline{w(z_0)} \\ F_{\bar{z}}(z_0) & F(z_0) & \overline{F(z_0)} \\ G_{\bar{z}}(z_0) & G(z_0) & \overline{G(z_0)} \end{vmatrix} = 0, \tag{2.5}$$

and if (2.2) exists, then

$$\dot{w}(z_0) = \frac{\begin{vmatrix} w_z(z_0) & w(z_0) & \overline{w(z_0)} \\ F_z(z_0) & F(z_0) & \overline{F(z_0)} \\ G_z(z_0) & G(z_0) & \overline{G(z_0)} \end{vmatrix}}{\begin{vmatrix} F(z_0) & \overline{F(z_0)} \\ G(z_0) & \overline{G(z_0)} \end{vmatrix}}. \tag{2.6}$$

The following expressions are known as *characteristic coefficients* of the pair (F, G):

$$a_{(F,G)} = -\frac{\overline{F}G_{\overline{z}} - F_{\overline{z}}\overline{G}}{F\overline{G} - \overline{F}G}, \qquad b_{(F,G)} = \frac{FG_{\overline{z}} - F_{\overline{z}}G}{F\overline{G} - \overline{F}G},$$

$$A_{(F,G)} = -\frac{\overline{F}G_z - F_z\overline{G}}{F\overline{G} - \overline{F}G}, \qquad B_{(F,G)} = \frac{FG_z - F_zG}{F\overline{G} - \overline{F}G}.$$

Equations (2.5) and (2.6) can be rewritten in the form

$$w_{\overline{z}} = a_{(F,G)}w + b_{(F,G)}\overline{w} \tag{2.7}$$

and

$$\dot{w} = w_z - A_{(F,G)}w - B_{(F,G)}\overline{w}. \tag{2.8}$$

Thus the following theorem is valid.

Theorem 3. *If $\dot{w}(z_0)$ exists, then at z_0, w_z and $w_{\overline{z}}$ exist and equations (2.7), (2.8) hold. If w_z and $w_{\overline{z}}$ exist and are continuous in some neighborhood of z_0, and if (2.7) holds at z_0, then $\dot{w}(z_0)$ exists, and (2.8) holds.*

Equation (2.7) is called a *Vekua equation* (sometimes, Carleman-Vekua equation).

Note that F and G possess (F, G)-derivatives, $\dot{F} \equiv \dot{G} \equiv 0$ and the following equalities which determine the characteristic coefficients uniquely, are valid:

$$F_{\overline{z}} = a_{(F,G)}F + b_{(F,G)}\overline{F}, \quad G_{\overline{z}} = a_{(F,G)}G + b_{(F,G)}\overline{G},$$

$$F_z = A_{(F,G)}F + B_{(F,G)}\overline{F}, \quad G_z = A_{(F,G)}G + B_{(F,G)}\overline{G}.$$

The Vekua equation (2.7) represents a generalization of the Cauchy-Riemann system and the main object of study of pseudoanalytic function theory.

2.2 Pseudoanalytic functions

Definition 4. A function w will be called (F, G)-*pseudoanalytic of the first kind* in a domain Ω (or, simply, *pseudoanalytic*, if there is no danger of confusion) if $\dot{w}$ exists everywhere in Ω.

In view of the condition $\text{Im}(\overline{F}G) > 0$ in Ω, every function w in a domain of interest admits the unique representation

$$w = \varphi F + \psi G$$

where φ and ψ are real-valued. Set

$$\omega = \varphi + i\psi.$$

The correspondence between w and ω is one-to-one. We denote it by writing

$$w =^* \omega, \qquad \omega =_* w \qquad (\text{mod } F, G).$$

Note that

$$_*(\lambda w_1 + \mu w_2) = \lambda(_*w_1) + \mu(_*w_2)$$

for any real λ and μ, $_*0 = 0$, $_*F = 1$, $_*G = i$ and that in every closed domain $\overline{\Omega}$,

$$0 < \frac{1}{K} \le \left| \frac{_*w(z)}{w(z)} \right| \le K$$

where the constant K depends only on (F, G) and the domain Ω.

Definition 5. If w is (F, G)-pseudoanalytic of the first kind, the function $\omega =_* w$ is called (F, G)-pseudoanalytic of the second kind.

In the case of analytic functions, F and G can be chosen as $F \equiv 1$, $G \equiv i$, and hence w coincides with ω.

The following theorem gives us an equation for pseudoanalytic functions of the second kind as well as a very useful and simple representation for the (F, G)-derivative of a pseudoanalytic function of the first kind.

Theorem 6. *A function $\omega = \varphi + i\psi$ is (F, G)-pseudoanalytic of the second kind if and only if φ and ψ possess continuous partial derivatives, and*

$$\varphi_{\bar{z}}F + \psi_{\bar{z}}G = 0. \tag{2.9}$$

If this condition is satisfied, then setting $w =^* \omega$ we have that

$$\dot{w} = \varphi_z F + \psi_z G. \tag{2.10}$$

Proof. Consider the function

$$W(z) = w(z) - \lambda_0 F(z) - \mu_0 G(z)$$

where $\lambda_0 = \varphi(z_0)$ and $\mu_0 = \psi(z_0)$, so that

$$w(z_0) = \lambda_0 F(z_0) + \mu_0 G(z_0).$$

Note that

$$W(z) = (\varphi(z) - \varphi(z_0))F(z) + (\psi(z) - \psi(z_0))G(z).$$

Then

$$W_{\bar{z}}(z) = \varphi_{\bar{z}}(z)F(z) + (\varphi(z) - \varphi(z_0))F_{\bar{z}}(z) + \psi_{\bar{z}}(z)G(z) + (\psi(z) - \psi(z_0))G_{\bar{z}}(z)$$

and hence at z_0 we have

$$W_{\bar{z}}(z_0) = \varphi_{\bar{z}}(z_0)F(z_0) + \psi_{\bar{z}}(z_0)G(z_0).$$

In a similar way we obtain that

$$W_z(z_0) = \varphi_z(z_0)F(z_0) + \psi_z(z_0)G(z_0).$$

As the Vekua equation

$$w_{\bar{z}} = aw + b\overline{w} \tag{2.11}$$

at z_0 is equivalent to the Cauchy-Riemann condition $W_{\bar{z}}(z_0) = 0$, we obtain that (2.11) is equivalent to (2.9), and due to the equality $\dot{w}(z_0) = W'(z_0)$ we obtain (2.10). $\qquad\square$

Example 7. Let f be a real-valued positive function. Consider $F = f$, $G = i/f$. Then equation (2.9) becomes

$$\varphi_{\bar{z}}f + \psi_{\bar{z}}\frac{i}{f} = 0$$

which is equivalent to the system

$$f\varphi_x - \frac{1}{f}\psi_y = 0, \qquad f\varphi_y + \frac{1}{f}\psi_x = 0.$$

This system can be written in the form of a generalized Cauchy-Riemann system with a "weight",

$$\varphi_x = \frac{1}{f^2}\psi_y, \qquad \varphi_y = -\frac{1}{f^2}\psi_x.$$

Note that the system

$$u_x = \frac{1}{p}v_y, \qquad u_y = -\frac{1}{p}v_x$$

where p is a given positive function of x and y is quite well known [103], and complex functions $u + iv$ satisfying it are called p-analytic.

Thus, $(f, i/f)$-pseudoanalytic functions of the second kind are f^2-analytic.

2.3 Derivatives and integrals of pseudoanalytic functions

2.3.1 Equivalent generating pairs

Definition 8. Two generating pairs (F, G) and $(\widetilde{F}, \widetilde{G})$ are called equivalent if

$$\widetilde{F} = a_{11}F + a_{12}G \quad \text{and} \quad \widetilde{G} = a_{21}F + a_{22}G$$

where a_{ij} are real constants.

The following theorem we give without proof for which we refer to [13].

Theorem 9.

1. *Two generating pairs defined in the same domain are equivalent if and only if they have the same characteristic coefficients.*
2. *If (F, G) and $(\widetilde{F}, \widetilde{G})$ are equivalent, then every $(\widetilde{F}, \widetilde{G})$-pseudoanalytic function of the first kind is (F, G)-pseudoanalytic of the first kind and*

$$\frac{d_{(F,G)}w}{dz} = \frac{d_{(\widetilde{F},\widetilde{G})}w}{dz}.$$

2.3.2   Vekua's equation for (F, G)-derivatives

A complex derivative of an analytic function is of course again an analytic function and both satisfy the Cauchy-Riemann system. The situation is different in the case of pseudoanalytic functions. The (F, G)-derivative of an (F, G)-pseudoanalytic function in general is no longer (F, G)-pseudoanalytic. Instead, it satisfies another Vekua equation, corresponding to another generating pair which is determined as follows.

Definition 10. Let (F, G) and (F_1, G_1) be two generating pairs in Ω. (F_1, G_1) is called successor of (F, G) and (F, G) is called predecessor of (F_1, G_1) if

$$a_{(F_1, G_1)} = a_{(F,G)} \qquad \text{and} \qquad b_{(F_1, G_1)} = -B_{(F,G)}.$$

The importance of this definition becomes obvious from the following statement.

Theorem 11. *Let w be an (F, G)-pseudoanalytic function and let (F_1, G_1) be a successor of (F, G). Then*

$$\dot{w} = \frac{d_{(F,G)}w}{dz}$$

is an (F_1, G_1)-pseudoanalytic function.

Proof. Set $\omega =_* w = \varphi + i\psi$. Then

$$\dot{w} = \varphi_z F + \psi_z G \tag{2.12}$$

and

$$\varphi_{\overline{z}} F + \psi_{\overline{z}} G = 0,$$

so that

$$\varphi_z \overline{F} + \psi_z \overline{G} = 0. \tag{2.13}$$

Solving (2.12) and (2.13) we get

$$\varphi_z = \frac{\overline{G}\dot{w}}{F\overline{G} - \overline{F}G} \qquad \text{and} \qquad \psi_z = -\frac{\overline{F}\dot{w}}{F\overline{G} - \overline{F}G}. \tag{2.14}$$

Note that

$$\varphi_{z\overline{z}}F + \psi_{z\overline{z}}G + \varphi_{\overline{z}}F_z + \psi_{\overline{z}}G_z = 0.$$

Consider

$$\begin{aligned}(\mathring{w})_{\overline{z}} &= \varphi_{z\overline{z}}F + \psi_{z\overline{z}}G + \varphi_z F_{\overline{z}} + \psi_z G_{\overline{z}} \\ &= \varphi_z F_{\overline{z}} + \psi_z G_{\overline{z}} - \varphi_{\overline{z}}F_z - \psi_{\overline{z}}G_z \\ &= \varphi_z F_{\overline{z}} + \psi_z G_{\overline{z}} - \left(\overline{\varphi_z \overline{F}_{\overline{z}} + \psi_z \overline{G}_{\overline{z}}}\right).\end{aligned}$$

Substituting (2.14) into this equality gives

$$\begin{aligned}(\mathring{w})_{\overline{z}} &= \frac{\overline{G}F_{\overline{z}} - \overline{F}G_{\overline{z}}}{F\overline{G} - \overline{F}G}\mathring{w} - \left(\overline{\frac{\overline{G}F_{\overline{z}} - \overline{F}G_{\overline{z}}}{F\overline{G} - \overline{F}G}}\right)\overline{\mathring{w}} \\ &= a\mathring{w} + \frac{GF_z - FG_z}{F\overline{G} - \overline{F}G}\overline{\mathring{w}} \\ &= a\mathring{w} - B\overline{\mathring{w}}.\end{aligned}$$ $\square$

Thus, $\mathring{w}$ is a solution of the Vekua equation

$$(\mathring{w})_{\overline{z}} = a\mathring{w} - B\overline{\mathring{w}}.$$

In order to introduce the notion of pseudoanalytic derivatives of arbitrary order the following very important definition is necessary.

Definition 12. A sequence of generating pairs $\{(F_m, G_m)\}$, $m = 0, \pm1, \pm2, \ldots$, is called a generating sequence if (F_{m+1}, G_{m+1}) is a successor of (F_m, G_m). If $(F_0, G_0) = (F, G)$, we say that (F, G) is embedded in $\{(F_m, G_m)\}$.

Theorem 13. *Let (F, G) be a generating pair in Ω. Let Ω_1 be a bounded domain, $\overline{\Omega}_1 \subset \Omega$. Then (F, G) can be embedded in a generating sequence in Ω_1.*

For the proof we refer to [13].

Definition 14. A generating sequence $\{(F_m, G_m)\}$ is said to have period $\mu > 0$ if $(F_{m+\mu}, G_{m+\mu})$ is equivalent to (F_m, G_m), that is their characteristic coefficients coincide.

Let W be an (F, G)-pseudoanalytic function. Using a generating sequence in which (F, G) is embedded we can define the higher derivatives of W by the recursion formula

$$W^{[0]} = W; \qquad W^{[m+1]} = \frac{d_{(F_m, G_m)}W^{[m]}}{dz}, \quad m = 0, 1, \ldots.$$

2.3.3   Integration

In the proof of Theorem 11 we obtained a couple of auxiliary relations (2.14):

$$\varphi_z = \frac{\overline{G}\dot{w}}{F\overline{G} - \overline{F}G} \qquad \text{and} \qquad \psi_z = -\frac{\overline{F}\dot{w}}{F\overline{G} - \overline{F}G}. \tag{2.15}$$

Now if we want to recover φ and ψ (and hence ω and w) from $\dot{w}$ we should integrate the expressions in (2.15). Let us consider this question in detail and introduce some useful notation.

Consider the equation

$$\varphi_z = \Phi \tag{2.16}$$

first in a whole complex plane or in a convex domain, where Φ is a complex-valued function $\Phi = \Phi_1 + i\Phi_2$ and φ is real-valued. It is easy to see that as this equation is equivalent to the system

$$\varphi_x = 2\Phi_1 \qquad \text{and} \qquad \varphi_y = -2\Phi_2,$$

it has a solution if only the following compatibility condition is satisfied:

$$\partial_y \Phi_1 + \partial_x \Phi_2 = 0. \tag{2.17}$$

If this condition is fulfilled, one can reconstruct φ up to an arbitrary real constant in the following way:

$$\varphi(x,y) = 2 \left(\int_{x_0}^{x} \Phi_1(\eta, y) d\eta - \int_{y_0}^{y} \Phi_2(x_0, \xi) d\xi \right) + c$$

where (x_0, y_0) is an arbitrary fixed point in the domain of interest. Note that this formula can be easily extended to any simply connected domain by considering the integral along an arbitrary rectifiable curve Γ leading from (x_0, y_0) to (x, y),

$$\varphi(x,y) = 2 \left(\int_{\Gamma} \Phi_1 dx - \Phi_2 dy \right) + c. \tag{2.18}$$

By A we denote the integral operator in (2.18):

$$A[\Phi](x,y) = 2 \left(\int_{\Gamma} \Phi_1 dx - \Phi_2 dy \right).$$

Notice that this expression can also be written as

$$A[\Phi](x,y) = 2 \operatorname{Re} \int_{\Gamma} (\Phi_1 + i\Phi_2) (dx + idy) = 2 \operatorname{Re} \int_{\Gamma} \Phi dz. \tag{2.19}$$

In a similar way we introduce the integral operator

$$\overline{A}[\Phi](\tilde{x}, y) = 2 \left(\int_{\Gamma} \Phi_1 dx + \Phi_2 dy \right)$$

corresponding to the operator $\partial_{\bar{z}}$ and applied to complex functions whose real and imaginary parts satisfy the condition

$$\partial_y \Phi_1 - \partial_x \Phi_2 = 0.$$

Returning to equalities (2.15) we see that to recover φ and ψ up to arbitrary real constants one can apply the operator A to the expressions on the right-hand side. Thus, we have

$$\varphi = A\left[\frac{\overline{G}\dot{w}}{F\overline{G} - \overline{F}G}\right] \quad \text{and} \quad \psi = -A\left[\frac{\overline{F}\dot{w}}{F\overline{G} - \overline{F}G}\right], \qquad (2.20)$$

and hence

$$\omega = A\left[\frac{\overline{G}\dot{w}}{F\overline{G} - \overline{F}G}\right] - iA\left[\frac{\overline{F}\dot{w}}{F\overline{G} - \overline{F}G}\right]$$

up to an additive complex constant which is the value of ω at $z_0 = (x_0, y_0)$, and

$$w = F \cdot A\left[\frac{\overline{G}\dot{w}}{F\overline{G} - \overline{F}G}\right] - G \cdot A\left[\frac{\overline{F}\dot{w}}{F\overline{G} - \overline{F}G}\right]$$

up to an additive term $c_1 F + c_2 G$ where c_1 and c_2 are arbitrary real constants. Fixing the value of w at $z_0 = (x_0, y_0)$ we find that $c_1 = \varphi(z_0)$ and $c_2 = \psi(z_0)$.

Taking into account (2.19) the last two equalities can be written in the following form, preferred in [13]:

$$\omega = \text{Re} \int_\Gamma \frac{2\overline{G}\dot{w}dz}{F\overline{G} - \overline{F}G} - i\,\text{Re} \int_\Gamma \frac{2\overline{F}\dot{w}dz}{F\overline{G} - \overline{F}G}$$

and

$$w = F\,\text{Re} \int_\Gamma \frac{2\overline{G}\dot{w}dz}{F\overline{G} - \overline{F}G} - G\,\text{Re} \int_\Gamma \frac{2\overline{F}\dot{w}dz}{F\overline{G} - \overline{F}G}.$$

These formulas lead naturally to the following definitions introduced in [13].

Definition 15. Let (F, G) be a generating pair. Its adjoint generating pair $(F, G)^* = (F^*, G^*)$ is defined by the formulas

$$F^* = -\frac{2\overline{F}}{F\overline{G} - \overline{F}G}, \qquad G^* = \frac{2\overline{G}}{F\overline{G} - \overline{F}G}.$$

Definition 16. The (F, G)-*-integral is defined by the equality

$$*\int_\Gamma W d_{(F,G)}z = \text{Re} \int_\Gamma G^* W dz + i\,\text{Re} \int_\Gamma F^* W dz$$

and the (F, G)-integral is defined as

$$\int_\Gamma W d_{(F,G)}z = F(z_1)\,\text{Re} \int_\Gamma G^* W dz + G(z_1)\,\text{Re} \int_\Gamma F^* W dz \qquad (2.21)$$

where Γ is a rectifiable curve leading from z_0 to z_1.

Definition 17. A continuous function W defined in a domain Ω will be called (F, G)-integrable if for every closed curve Γ lying in a simply connected subdomain of Ω,

$$\oint_{\Gamma} W d_{(F,G)} z = 0.$$

Theorem 18. *An (F, G)-derivative $\dot{w}$ of an (F, G)-pseudoanalytic function w is (F, G)-integrable.*

Proof. It follows from the path-independence of the integrals in (2.20).   □

Theorem 19. *Let $\dot{w}$ be an (F, G)-derivative of an (F, G)-pseudoanalytic function w in a simply connected domain Ω and $\Gamma \subset \Omega$ be a rectifiable curve leading from z_0 to z. Then the following equalities are valid:*

$$* \int_{\Gamma} \dot{w} d_{(F,G)} z = \omega(z) - \omega(z_0), \qquad \omega =_* w \quad (\mathrm{mod}\ F, G),$$

$$\int_{\Gamma} \dot{w} d_{(F,G)} z = w(z) - \varphi(z_0) F(z) - \psi(z_0) G(z).$$

Proof. We obtained it earlier as a corollary of relations (2.20).   □

The integral $\int_{z_0}^{z_1} \dot{w} d_{(F,G)} z$ is called an (F, G)-antiderivative of $\dot{w}$.

Theorem 20. *Let (F, G) be a predecessor of (F_1, G_1). A continuous function is (F_1, G_1)-pseudoanalytic if and only if it is (F, G)-integrable.*

Theorem 21. *If W is a continuous function defined in a simply connected domain Ω, and if W is (F, G)-integrable, then there exists an (F, G)-pseudoanalytic function w in Ω, such that*

$$W(z) = \frac{d_{(F,G)} w(z)}{dz}.$$

Proof. Under the hypotheses of the theorem the function

$$\omega = \varphi + i\psi = * \int_{\Gamma} W d_{(F,G)} z$$

is well defined and possesses continuous partial derivatives (z_0 being any fixed point in Ω). In fact

$$\varphi = \mathrm{Re} \int_{\Gamma} G^* W dz = \int_{\Gamma} \frac{\overline{G} W dz - G \overline{W} d\bar{z}}{F\overline{G} - \overline{F}G}$$

and

$$\psi = \mathrm{Re} \int_{\Gamma} F^* W dz = - \int_{\Gamma} \frac{\overline{F} W dz - F \overline{W} d\bar{z}}{F\overline{G} - \overline{F}G}$$

from where it follows that

$$\varphi_z = \frac{\overline{G}W}{F\overline{G} - \overline{F}G}, \qquad \psi_z = -\frac{\overline{F}W}{F\overline{G} - \overline{F}G}$$

and

$$\varphi_{\bar{z}} = -\frac{G\overline{W}}{F\overline{G} - \overline{F}G}, \qquad \psi_{\bar{z}} = \frac{F\overline{W}}{F\overline{G} - \overline{F}G}.$$

Hence

$$\varphi_{\bar{z}}F + \psi_{\bar{z}}G = 0 \qquad \text{and} \qquad \varphi_z F + \psi_z G = W.$$

From Theorem 6 we obtain that $w = \varphi F + \psi G$ is (F, G)-pseudoanalytic, and that $\dot{w} = W$. $\qquad\square$

Remark 22. Theorem 21 holds also for multiply-connected domains, except that w may be multiple-valued.

Let us formulate an auxiliary fact concerning the adjoint generating pair introduced in Definition 15.

Theorem 23.

1. $(F, G)^{**} = (F, G)$.

2. *The following relations between characteristic coefficients hold:*

$$a_{(F^*,G^*)} = -a_{(F,G)}, \qquad A_{(F^*,G^*)} = -A_{(F,G)},$$
$$b_{(F^*,G^*)} = -\overline{B_{(F,G)}}, \qquad B_{(F^*,G^*)} = -\overline{b_{(F,G)}}.$$

Proof. The proof is straightforward. $\qquad\square$

Now we prove a statement which is converse with respect to Theorem 11.

Theorem 24. *Let (F_1, G_1) be a successor of (F, G), and let W be an (F_1, G_1)-pseudoanalytic function. Then W is (F, G)-integrable and hence an (F, G)-derivative of (a not necessarily single-valued) (F, G)-pseudoanalytic function.*

Proof. Due to Theorem 21 it is sufficient to prove that if Ω is a regular domain[2], and $\overline{\Omega}$ lies within the domain of definition of W, then

$$*\int_{\partial\Omega} W d_{(F,G)}z = 0. \tag{2.22}$$

Let (F^*, G^*) be the adjoint of (F, G). Then

$$\text{Re}\left(*\int_{\partial\Omega} W d_{(F,G)}z\right) = \text{Re}\int_{\partial\Omega} G^* W dz$$

---

[2]If Ω is bounded and $\partial\Omega$ consists of a finite number of piecewise continuously differentiable simple closed Jordan curves, Ω will be called **regular**.

and

$$\text{Im}\left(*\int_{\partial\Omega} W d_{(F,G)}z\right) = \text{Re}\int_{\partial\Omega} F^*W dz$$

for every W. By Theorem 23

$$F^*_{\bar{z}} = -aF^* - \overline{BF^*} \qquad \text{and} \qquad G^*_{\bar{z}} = -aG^* - \overline{BG^*}$$

and by hypothesis

$$W_{\bar{z}} = aW - B\overline{W}$$

where a, b, and B are the characteristic coefficients of (F, G).

Here we are going to make use of one of the complex versions of the Green-Gauss integral theorem (see, e.g., [117, Sect. 3.2]) which establishes that for a regular domain Ω and any complex, continuously differentiable with respect to x and y, function g defined in $\overline{\Omega}$ the following equality holds:

$$\int_{\Omega} g_{\bar{z}} dx dy = \frac{1}{2i}\int_{\partial\Omega} g dz.$$

We have that the integral

$$\int_{\partial\Omega} F^*W dz = 2i\int_{\Omega} (F^*W)_{\bar{z}} dx dy$$

$$= 2i\int_{\Omega}\left(-aF^*W - \overline{BF^*}W + F^*aW - F^*B\overline{W}\right) dx dy$$

$$= -4i\int_{\Omega} \text{Re}\left(F^*B\overline{W}\right) dx dy$$

is a pure imaginary number and hence $\text{Re}\int_{\partial\Omega} F^*W dz = 0$. The same reasoning shows that $\text{Re}\int_{\partial\Omega} G^*W dz = 0$ so that (2.22) is valid. $\square$

Chapter 3

Solutions of Second-order Elliptic Equations as Real Components of Complex Pseudoanalytic Functions

3.1 Factorization of the stationary Schrödinger operator

It is well known that if f_0 is a nonvanishing particular solution of the one-dimensional stationary Schrödinger equation

$$\left(-\frac{d^2}{dx^2} + \nu(x)\right) f(x) = 0,$$

then the Schrödinger operator can be factorized as

$$\frac{d^2}{dx^2} - \nu(x) = \left(\frac{d}{dx} + \frac{f_0'}{f_0}\right)\left(\frac{d}{dx} - \frac{f_0'}{f_0}\right).$$

We start with a generalization of this result onto a two-dimensional situation. Consider the two-dimensional stationary Schrödinger equation

$$(-\Delta + \nu) f = 0 \tag{3.1}$$

in some domain $\Omega \subset \mathbf{R}^2$, where $\Delta = \frac{\partial^2}{\partial x^2} + \frac{\partial^2}{\partial y^2}$, ν and f are real-valued functions. We assume that f is a twice-continuously differentiable function. By C we denote the complex conjugation operator.

Theorem 25 ([69]). *Let f be a positive in Ω particular solution of* (3.1). *Then for any real-valued function $\varphi \in C^2(\Omega)$ the following equalities hold:*

$$\frac{1}{4}(\Delta - \nu)\varphi = \left(\partial_{\bar{z}} + \frac{f_z}{f}C\right)\left(\partial_z - \frac{f_z}{f}C\right)\varphi = \left(\partial_z + \frac{f_{\bar{z}}}{f}C\right)\left(\partial_{\bar{z}} - \frac{f_{\bar{z}}}{f}C\right)\varphi.$$
$$\tag{3.2}$$

Proof. Consider

$$\left(\partial_{\bar{z}} + \frac{f_z}{f}C\right)\left(\partial_z - \frac{f_z}{f}C\right)\varphi = \frac{1}{4}\Delta\varphi - \frac{|\partial_z f|^2}{f^2}\varphi - \partial_{\bar{z}}\left(\frac{\partial_z f}{f}\right)\varphi$$

$$= \frac{1}{4}\left(\Delta\varphi - \frac{\Delta f}{f}\varphi\right) = \frac{1}{4}\left(\Delta - \nu\right)\varphi. \qquad (3.3)$$

Thus, we have the first equality in (3.2). Now application of C to both sides of (3.3) gives us the second equality in (3.2). $\qquad\qquad\square$

The operator $\partial_z - \frac{f_z}{f}I$, where I is the identity operator, can be represented in the form

$$\partial_z - \frac{f_z}{f}I = f\partial_z f^{-1}I.$$

Let us introduce the notation $P = f\partial_z f^{-1}I$. Due to Theorem 25, if f is a positive solution of (3.1), the operator P transforms real-valued solutions of (3.1) into solutions of the Vekua equation

$$\left(\partial_{\bar{z}} + \frac{f_z}{f}C\right)w = 0. \qquad (3.4)$$

Consider the operator $S = fAf^{-1}I$ applicable to any complex-valued function w such that $\Phi = f^{-1}w$ satisfies condition (2.17). Then it is clear that for such w we have that $PSw = w$.

Proposition 26. [70] *Let f be a positive particular solution of (3.1) and w be a solution of (3.4). Then the real-valued function $g = Sw$ is a solution of (3.1).*

Proof. First of all let us check that the function $\Phi = w/f$ satisfies (2.17). Let $u = \operatorname{Re} w$ and $v = \operatorname{Im} w$. Consider

$$\partial_y \Phi_1 + \partial_x \Phi_2 = \frac{1}{f}\left(\left(\partial_y u + \partial_x v\right) - \left(\frac{\partial_y f}{f}u + \frac{\partial_x f}{f}v\right)\right). \qquad (3.5)$$

Note that equation (3.4) is equivalent to the system

$$\partial_x u - \partial_y v = -\frac{\partial_x f}{f}u + \frac{\partial_y f}{f}v, \qquad \partial_y u + \partial_x v = \frac{\partial_y f}{f}u + \frac{\partial_x f}{f}v$$

from which we obtain that expression (3.5) is zero. Thus the function Φ satisfies (2.17) and hence the real-valued function $\varphi = A[w/f]$ is well defined and satisfies the equation $\partial_z\varphi = w/f$.

Consider the expression

$$\partial_{\bar{z}}\partial_z(Sw) = \partial_{\bar{z}}\left((\partial_z f) A\left[\frac{w}{f}\right] + w\right)$$

$$= \left(\frac{1}{4}\Delta f\right)A\left[\frac{w}{f}\right] + (\partial_z f)\partial_{\bar{z}}A\left[\frac{w}{f}\right] - \frac{\partial_z f}{f}\overline{w}. \qquad (3.6)$$

For the expression $\partial_{\bar{z}} A[\frac{w}{f}]$ we have

$$\partial_{\bar{z}} A[\frac{w}{f}] = \partial_z A[\frac{w}{f}] + i\partial_y A[\frac{w}{f}]$$

$$= \frac{w}{f} - 2i\frac{v}{f} = \frac{\bar{w}}{f} \qquad (3.7)$$

where the following observation was used:

$$\partial_y A[\frac{u+iv}{f}](x,y) = 2\left(\int_{x_0}^x \partial_y\left(\frac{u(\eta,y)}{f(\eta,y)}\right)d\eta - \frac{v(x_0,y)}{f(x_0,y)}\right)$$

$$= -2\left(\int_{x_0}^x \partial_\eta\left(\frac{v(\eta,y)}{f(\eta,y)}\right)d\eta - \frac{v(x_0,y)}{f(x_0,y)}\right) = -\frac{2v(x,y)}{f(x,y)}.$$

Thus substitution of (3.7) into (3.6) gives us the equality

$$\Delta(Sw) = \nu f A[\frac{w}{f}] = \nu Sw. \qquad \square$$

Proposition 27. [70] *Let g be a real-valued solution of* (3.1)*. Then*

$$SPg = g + cf$$

where c is an arbitrary real constant.

Proof. Consider

$$SPg = f A\partial_z\left[\frac{g}{f}\right] = f\left(\frac{g}{f} + c\right) = g + cf. \qquad \square$$

Theorem 25 together with Proposition 26 show us that equation (3.1) is equivalent to the Vekua equation (3.4) in the following sense. Every solution of one of these equations can be transformed into a solution of the other equation and vice versa.

3.2 Factorization of the operator $\operatorname{div} p \operatorname{grad} + q$.

The following statement is known in the form of a substitution (see, e.g., [96]). Here we formulate it as an operational relation.

Proposition 28. *Let p and q be real-valued functions, $p \in C^2(\Omega)$ and $p \neq 0$ in Ω. Then*

$$\operatorname{div} p \operatorname{grad} + q = p^{1/2}(\Delta - r)p^{1/2} \qquad in \ \Omega, \qquad (3.8)$$

where

$$r = \frac{\Delta p^{1/2}}{p^{1/2}} - \frac{q}{p}.$$

Proof. The easily verified relation

$$\operatorname{div} p \operatorname{grad} = p^{1/2} \left(\Delta - \frac{\Delta p^{1/2}}{p^{1/2}} \right) p^{1/2} \tag{3.9}$$

is well known (see, e.g., [118]). Adding to both sides of (3.9) the term q (and representing it on the right-hand side as $p^{1/2} (q/p) p^{1/2}$) gives us (3.8). $\square$

The following statement is a generalization of Theorem 25.

Theorem 29 ([71]). *Let p and q be real-valued functions, $p \in C^2(\Omega)$ and $p \neq 0$ in Ω, u_0 be a positive particular solution of the equation*

$$(\operatorname{div} p \operatorname{grad} + q)u = 0 \qquad in \ \Omega. \tag{3.10}$$

Then for any real-valued twice-continuously differentiable function φ the following equality holds:

$$\frac{1}{4}(\operatorname{div} p \operatorname{grad} + q)\varphi = p^{1/2} \left(\partial_z + \frac{f_{\bar{z}}}{f}C \right) \left(\partial_{\bar{z}} - \frac{f_{\bar{z}}}{f}C \right) p^{1/2}\varphi, \tag{3.11}$$

where

$$f = p^{1/2} u_0. \tag{3.12}$$

Proof. This is based on (3.2). From (3.8) we have that if u_0 is a solution of (3.10) then the function (3.12) is a solution of the equation

$$(\Delta - r)f = 0. \tag{3.13}$$

Then combining (3.8) and (3.2) we obtain (3.11). $\square$

Remark 30. According to (3.9), $\Delta - r = f^{-1} \operatorname{div} f^2 \operatorname{grad} f^{-1}$ where f is a solution of (3.13). Then from (3.8) we have

$$\operatorname{div} p \operatorname{grad} + q = p^{1/2} f^{-1} \operatorname{div} f^2 \operatorname{grad} f^{-1} p^{1/2}. \tag{3.14}$$

Taking into account (3.12) we obtain

$$\operatorname{div} p \operatorname{grad} + q = u_0^{-1} \operatorname{div} p u_0^2 \operatorname{grad} u_0^{-1} \qquad in \ \Omega.$$

Remark 31. Let $q \equiv 0$. Then u_0 can be chosen as $u_0 \equiv 1$. Hence (3.11) gives us the equality

$$\frac{1}{4} \operatorname{div}(p \operatorname{grad} \varphi) = p^{1/2} \left(\partial_z + \frac{\partial_{\bar{z}} p^{1/2}}{p^{1/2}}C \right) \left(\partial_{\bar{z}} - \frac{\partial_{\bar{z}} p^{1/2}}{p^{1/2}}C \right) (p^{1/2}\varphi).$$

In what follows we suppose that in Ω there exists a positive particular solution of (3.10) which we denote by u_0.

Let f be a real function of x and y. Consider the Vekua equation

$$W_{\overline{z}} = \frac{f_{\overline{z}}}{f}\overline{W} \qquad \text{in } \Omega. \tag{3.15}$$

This equation plays a crucial role in all that follows, hence we will call it the *main Vekua equation*. The operator $\partial_{\overline{z}} - \frac{f_{\overline{z}}}{f}C$ corresponding to this equation appears in the factorization (3.11) as well as in (3.2).

Denote $W_1 = \operatorname{Re} W$ and $W_2 = \operatorname{Im} W$.

Remark 32. [69] Equation (3.15) can be written as

$$f\partial_{\overline{z}}(f^{-1}W_1) + if^{-1}\partial_{\overline{z}}(fW_2) = 0. \tag{3.16}$$

Theorem 33 ([71]). *Let $W = W_1 + iW_2$ be a solution of (3.15). Then $U = f^{-1}W_1$ is a solution of the conductivity equation*

$$\operatorname{div}(f^2 \nabla U) = 0 \qquad \text{in } \Omega, \tag{3.17}$$

and $V = fW_2$ is a solution of the associated conductivity equation

$$\operatorname{div}(f^{-2} \nabla V) = 0 \qquad \text{in } \Omega, \tag{3.18}$$

the function W_1 is a solution of the stationary Schrödinger equation

$$-\Delta W_1 + r_1 W_1 = 0 \qquad \text{in } \Omega \tag{3.19}$$

with $r_1 = \Delta f/f$, and W_2 is a solution of the associated stationary Schrödinger equation

$$-\Delta W_2 + r_2 W_2 = 0 \qquad \text{in } \Omega \tag{3.20}$$

where $r_2 = 2(\nabla f)^2/f^2 - r_1$ and $(\nabla f)^2 = f_x^2 + f_y^2$.

Proof. To prove the first part of the theorem we use the form of equation (3.15) given in Remark 32. Multiplying (3.16) by f and applying ∂_z gives

$$\partial_z \left(f^2 \partial_{\overline{z}} \left(f^{-1} W_1 \right) \right) + \frac{i}{4}\Delta \left(fW_2 \right) = 0$$

from where we have that $\operatorname{Re}\left(\partial_z \left(f^2 \partial_{\overline{z}} \left(f^{-1} W_1 \right) \right)\right) = 0$ which is equivalent to (3.17) where $U = f^{-1}W_1$.

Multiplying (3.16) by f^{-1} and applying ∂_z gives

$$\frac{1}{4}\Delta \left(f^{-1}W_1 \right) + i\partial_z \left(f^{-2}\partial_{\overline{z}} \left(fW_2 \right) \right) = 0$$

from where we have that $\operatorname{Re}\left(\partial_z \left(f^{-2}\partial_{\overline{z}} \left(fW_2 \right) \right)\right) = 0$ which is equivalent to (3.18) where $V = fW_2$.

From (3.9) we have

$$(\Delta - r_1)\, W_1 = f^{-1}\, \mathrm{div}(f^2 \nabla \left(f^{-1} W_1 \right)).$$

Hence from the just proven equation (3.17) we obtain that W_1 is a solution of (3.19).

In order to obtain equation (3.20) for W_2 it should be noticed that

$$f\, \mathrm{div}(f^{-2} \nabla (f W_2)) = (\Delta - r_2)\, W_2. \qquad \square$$

Remark 34. Observe that the pair of functions

$$F = f \quad \text{and} \quad G = \frac{i}{f} \tag{3.21}$$

are solutions of (3.15) and represent a generating pair corresponding to the Vekua equation (3.15). This allows us to rewrite (3.15) in the form of an equation for pseudoanalytic functions of second kind (equation (2.9))

$$\varphi_{\overline{z}} f + \psi_{\overline{z}} \frac{i}{f} = 0, \tag{3.22}$$

where φ and ψ are real-valued functions. If φ and ψ satisfy (3.22) then $W = \varphi f + \psi \frac{i}{f}$ is a solution of (3.15) and vice versa.

Denote $w = \varphi + \psi i$. Then from (3.22) we have

$$(w + \overline{w})_{\overline{z}} f + (w - \overline{w})_{\overline{z}} \frac{1}{f} = 0,$$

which is equivalent to the equation

$$w_{\overline{z}} = \frac{1 - f^2}{1 + f^2} \overline{w}_{\overline{z}}. \tag{3.23}$$

The relation between (3.23) and (3.17), (3.18) was observed in [3] and turned out to be essential for solving the Calderón problem in the plane.

Theorem 35 ([71]). *Let $W = W_1 + i W_2$ be a solution of (3.15). Assume that $f = p^{1/2} u_0$, where u_0 is a positive solution of (3.10) in Ω. Then $u = p^{-1/2} W_1$ is a solution of (3.10) in Ω, and $v = p^{1/2} W_2$ is a solution of the equation*

$$\left(\mathrm{div}\, \frac{1}{p}\, \mathrm{grad} + q_1 \right) v = 0 \qquad \text{in } \Omega, \tag{3.24}$$

where

$$q_1 = -\frac{1}{p} \left(\frac{q}{p} + 2 \left\langle \frac{\nabla p}{p}, \frac{\nabla u_0}{u_0} \right\rangle + 2 \left(\frac{\nabla u_0}{u_0} \right)^2 \right). \tag{3.25}$$

Proof. According to Theorem 33, the function $f^{-1}W_1$ is a solution of (3.17). From (3.14) we have that

$$p^{-1/2}\left(\operatorname{div}p\operatorname{grad}+q\right)(p^{-1/2}W_1) = f^{-1}\operatorname{div}(f^2\nabla(f^{-1}W_1))$$

from which we obtain that $u = p^{-1/2}W_1$ is a solution of (3.10).

In order to obtain the second assertion of the theorem, let us show that

$$p^{1/2}\left(\operatorname{div}\frac{1}{p}\operatorname{grad}+q_1\right)(p^{1/2}\varphi) = f\operatorname{div}(f^{-2}\nabla(f\varphi))$$

for any real-valued $\varphi \in C^2(\Omega)$. According to (3.9),

$$f\operatorname{div}(f^{-2}\nabla(f\varphi)) = \left(\Delta - \frac{\Delta f^{-1}}{f^{-1}}\right)\varphi = (\Delta - r_2)\varphi.$$

Straightforward calculation gives us the equality

$$\frac{\Delta f^{-1}}{f^{-1}} = \frac{3}{4}\left(\frac{\nabla p}{p}\right)^2 - \frac{1}{2}\frac{\Delta p}{p} + \left\langle\frac{\nabla p}{p},\frac{\nabla u_0}{u_0}\right\rangle - \frac{\Delta u_0}{u_0} + 2\left(\frac{\nabla u_0}{u_0}\right)^2.$$

From the condition that u_0 is a solution of (3.10) we obtain the equality

$$-\frac{\Delta u_0}{u_0} = \frac{q}{p} + \left\langle\frac{\nabla p}{p},\frac{\nabla u_0}{u_0}\right\rangle.$$

Thus,

$$\frac{\Delta f^{-1}}{f^{-1}} = \frac{3}{4}\left(\frac{\nabla p}{p}\right)^2 - \frac{1}{2}\frac{\Delta p}{p} + 2\left\langle\frac{\nabla p}{p},\frac{\nabla u_0}{u_0}\right\rangle + \frac{q}{p} + 2\left(\frac{\nabla u_0}{u_0}\right)^2.$$

Notice that

$$\frac{\Delta p^{-1/2}}{p^{-1/2}} = \frac{3}{4}\left(\frac{\nabla p}{p}\right)^2 - \frac{1}{2}\frac{\Delta p}{p}.$$

Then

$$\frac{\Delta f^{-1}}{f^{-1}} = \frac{\Delta p^{-1/2}}{p^{-1/2}} + 2\left\langle\frac{\nabla p}{p},\frac{\nabla u_0}{u_0}\right\rangle + \frac{q}{p} + 2\left(\frac{\nabla u_0}{u_0}\right)^2.$$

Now taking q_1 in the form (3.25) we obtain the result from (3.8).   □

3.3 Conjugate metaharmonic functions

Theorems 33 and 35 show us that as much as real and imaginary parts of a complex analytic function are harmonic functions, the real and imaginary parts of a solution of the main Vekua equation (3.15) are solutions of associated stationary Schrödinger equations, being also related to conductivity equations as well as to more general elliptic equations (3.10) and (3.24). The following natural question arises then. We know that given an arbitrary real-valued harmonic function in a simply connected domain, a conjugate harmonic function can be constructed explicitly such that the obtained pair of harmonic functions represent the real and imaginary parts of a complex analytic function. This corresponds to the more general fact for solutions of associated stationary Schrödinger equations (which, slightly generalizing the definition of I.N. Vekua, we call metaharmonic functions) and of other aforementioned elliptic equations. The precise result for the Schrödinger equations is given in the following theorem.

Theorem 36 ([69]). *Let W_1 be a real-valued solution of* (3.19) *in a simply connected domain Ω. Then the real-valued function W_2, a solution of* (3.20) *such that $W = W_1 + iW_2$ is a solution of* (3.15), *is constructed according to the formula*

$$W_2 = f^{-1}\overline{A}(if^2\partial_{\overline{z}}(f^{-1}W_1)). \tag{3.26}$$

Given a solution W_2 of (3.20), *the corresponding solution W_1 of* (3.19) *such that $W = W_1 + iW_2$ is a solution of* (3.15), *is constructed as*

$$W_1 = -f\overline{A}(if^{-2}\partial_{\overline{z}}(fW_2)). \tag{3.27}$$

Proof. Consider equation (3.15). Let $W = \phi f + i\psi/f$ be its solution. Then the equation

$$\psi_{\overline{z}} - if^2\phi_{\overline{z}} = 0 \tag{3.28}$$

is valid. Note that if $W_1 = \operatorname{Re} W$, then $\phi = W_1/f$. Given ϕ, ψ is easily found from (3.28):

$$\psi = \overline{A}(if^2\phi_{\overline{z}}).$$

It can be verified that the expression $\overline{A}(if^2\phi_{\overline{z}})$ makes sense, that is $\partial_x(f^2\phi_x) + \partial_y(f^2\phi_y) = 0$.

By Theorem 33 the function $W_2 = f^{-1}\psi$ is a solution of (3.20). Thus we obtain (3.26). Let us notice that as the operator $\overline{A}$ reconstructs the real function up to an arbitrary real constant, the function W_2 in the formula (3.26) is uniquely determined up to an additive term cf^{-1} where c is an arbitrary real constant.

Equation (3.27) is proved in a similar way. $\qquad\square$

Remark 37. When in (3.19) $r_1 \equiv 0$ and $f \equiv 1$, equalities (3.26) and (3.27) turn into the well-known formulas in complex analysis for constructing conjugate harmonic functions.

Corollary 38. [71] *Let U be a solution of* (3.17). *Then a solution V of* (3.18) *such that*

$$W = fU + if^{-1}V$$

is a solution of (3.15), *is constructed according to the formula*

$$V = \overline{A}(if^2U_{\overline{z}}). \tag{3.29}$$

Conversely, given a solution V of (3.18), the corresponding solution U of (3.17) can be constructed as

$$U = -\overline{A}(if^{-2}V_{\overline{z}}).$$

Proof. Consists in substitution of $W_1 = fU$ and of $W_2 = f^{-1}V$ into (3.26) and (3.27). $\qquad\square$

Corollary 39. [71] *Let $f = p^{1/2}u_0$, where u_0 is a positive solution of (3.10) in a simply connected domain Ω and u be a solution of (3.10). Then a solution v of (3.24) with q_1 defined by (3.25) such that $W = p^{1/2}u + ip^{-1/2}v$ is a solution of (3.15), is constructed according to the formula*

$$v = u_0^{-1}\overline{A}(ipu_0^2\partial_{\overline{z}}(u_0^{-1}u)).$$

Let v be a solution of (3.24), then the corresponding solution u of (3.10) such that $W = p^{1/2}u + ip^{-1/2}v$ is a solution of (3.15), is constructed according to the formula

$$u = -u_0\overline{A}(ip^{-1}u_0^{-2}\partial_{\overline{z}}(u_0v)).$$

Proof. Consists in substitution of $f = p^{1/2}u_0$, $W_1 = p^{1/2}u$ and $W_2 = p^{-1/2}v$ into (3.26) and (3.27). $\qquad\square$

3.4 The main Vekua equation

The results of the preceding section show us that the theory of the elliptic equation

$$(\operatorname{div} p \operatorname{grad} +q)u = 0 \tag{3.30}$$

is closely related to the equation (3.15):

$$W_{\overline{z}} = \frac{f_{\overline{z}}}{f}\overline{W}. \tag{3.31}$$

In fact if $f = p^{1/2}u_0$ where u_0 is a positive solution of (3.30) in a domain of interest, then for any solution u of (3.30) there exists a corresponding solution W of (3.31) such that

$$u = p^{-1/2}\operatorname{Re}W, \tag{3.32}$$

and vice versa, given a solution W of (3.31) the function (3.32) will be a solution of (3.30). As was pointed out in Remark 34, the pair of functions: $F = f$ and $G = \frac{i}{f}$ is a generating pair for this equation. Then the corresponding characteristic coefficients $A_{(F,G)}$ and $B_{(F,G)}$ have the form

$$A_{(F,G)} = 0, \qquad B_{(F,G)} = \frac{f_z}{f},$$

and the (F, G)-derivative according to (2.10) is defined as

$$\dot{W} = W_z - \frac{f_z}{f}\overline{W} = \left(\partial_z - \frac{f_z}{f}C\right)W.$$

Comparing $B_{(F,G)}$ with the coefficient in (3.4) and due to Theorem 11 we obtain the following statement.

Proposition 40. *Let W be a solution of* (3.31). *Then its (F, G)-derivative, the function $w = \dot{W}$ is a solution of* (3.4).

This result can be verified also by a direct substitution.

According to (2.21) and taking into account that

$$F^* = -if \quad \text{and} \quad G^* = 1/f,$$

the (F, G)-antiderivative has the form

$$\int_{z_0}^{z} w(\zeta)d_{(F,G)}\zeta = f(z)\,\text{Re}\int_{z_0}^{z}\frac{w(\zeta)}{f(\zeta)}d\zeta - \frac{i}{f(z)}\,\text{Re}\int_{z_0}^{z}if(\zeta)w(\zeta)d\zeta$$

$$= f(z)\,\text{Re}\int_{z_0}^{z}\frac{w(\zeta)}{f(\zeta)}d\zeta + \frac{i}{f(z)}\,\text{Im}\int_{z_0}^{z}f(\zeta)w(\zeta)d\zeta, \qquad (3.33)$$

and we obtain the following statement.

Proposition 41. *Let w be a solution of* (3.4). *Then the function*

$$W(z) = \int_{z_0}^{z} w(\zeta)d_{(F,G)}\zeta$$

is a solution of (3.31).

From the established relation between the main Vekua equation (3.31) and the Vekua equation (3.4) for solutions of the latter one, we obtain the following Cauchy integral theorem.

Theorem 42. *Let w be a solution of* (3.4) *in a domain Ω. Then for every closed curve Γ situated in a simply connected subdomain of Ω,*

$$\text{Re}\int_{\Gamma}\frac{w}{f}dz + i\,\text{Im}\int_{\Gamma}fwdz = 0. \qquad (3.34)$$

Proof. By Theorem 18 w is (F, G)-integrable. That is

$$\text{Im}\int_{\Gamma}\frac{w}{f}dz - i\,\text{Re}\int_{\Gamma}fwdz = 0. \qquad \square$$

3.5 Cauchy's integral theorem for the Schrödinger equation

Theorem 43 ([68] (Cauchy's integral theorem for the Schrödinger equation)). *Let f be a positive solution of (3.1) in a domain Ω and u be another arbitrary real-valued solution of (3.1) in Ω. Then for every closed curve Γ situated in a simply connected subdomain of Ω,*

$$\mathrm{Re} \int_{\Gamma} \partial_z \left(\frac{u}{f} \right) dz + i \,\mathrm{Im} \int_{\Gamma} f^2 \partial_z \left(\frac{u}{f} \right) dz = 0. \tag{3.35}$$

Proof. Substitution of $w = Pu = f \partial_z(u/f)$ (see Section 3.1) into (3.34) gives us the result. $\qquad \square$

Remark 44. This theorem is also valid when $f \equiv 1$ that is for u being a harmonic function. Then (3.35) turns into the equality $\int_{\Gamma} \partial_z u \, dz = 0$ which is obviously true because if u is harmonic, then $\partial_z u$ is analytic.

In Example 46 we will give a nontrivial example illustrating this theorem.

From Theorem 20 and Theorem 43 we obtain an analogue of the Morera theorem for the Schrödinger equation (3.1).

Theorem 45. *Let f be a positive particular solution of (3.1). A C^2-real-valued function u is a solution of (3.1) as well if (3.35) is valid for every closed curve Γ situated in a simply connected subdomain of Ω.*

Example 46. In order to illustrate the Cauchy integral theorem for the Schrödinger equation, let us consider the following two functions. Let $f(x, y) = e^{xy}$ and as u we choose the function e^{-xy}. Both f and u are solutions of (3.1) with the same potential $\nu(x, y) = x^2 + y^2$ in a whole plane. Thus we can apply Theorem 43 and consider Γ being, for example, a unit circle with centre at the origin. Then

$$\mathrm{Re} \int_{\Gamma} \partial_z \left(\frac{u}{f} \right) dz + i \,\mathrm{Im} \int_{\Gamma} f^2 \partial_z \left(\frac{u}{f} \right) dz$$

$$= \mathrm{Re} \int_{\Gamma} (-y + ix) e^{-2xy} dz + i \,\mathrm{Im} \int_{\Gamma} (-y + ix) dz$$

$$= \mathrm{Re} \int_0^{2\pi} i(-\sin\tau + i\cos\tau) e^{-2\cos\tau \sin\tau} d\tau + i \,\mathrm{Im} \int_0^{2\pi} i(-\sin\tau + i\cos\tau) d\tau$$

$$= - \int_0^{2\pi} \cos\tau \cdot e^{-2\cos\tau \sin\tau} d\tau - i \int_0^{2\pi} \sin\tau \, d\tau.$$

It is easy to see that both integrals are equal to zero.

3.6   *p*-analytic functions

Definition 47. A function $\Phi = u + iv$ of a complex variable $z = x + iy$ is said to be *p*-analytic in some domain Ω iff

$$u_x = \frac{1}{p}v_y, \qquad u_y = -\frac{1}{p}v_x \qquad \text{in } \Omega \tag{3.36}$$

where p is a given positive function of x and y which is supposed to be continuously differentiable.

The theory of *p*-analytic functions was presented in [103]. *p*-analytic functions in a certain sense represent a subclass of pseudoanalytic functions and it should be noticed that this subclass preserves some important properties of usual analytic functions which are not preserved by a too ample class of pseudoanalytic functions (corresponding details can be found in [103]). As we show here (see also Section 5.2) *p*-analytic functions are closely related to the solutions of the main Vekua equation. They have found numerous applications in elasticity theory (see, e.g., [2], [49]), in problems of fluid dynamics (see, e.g., [47], [48], [101], [102], [127]).

As was pointed out in Remark 34, the function $W = \phi f + i\psi/f$ is a solution of the main Vekua equation (3.31) if and only if ϕ and ψ satisfy the equation (3.22) which is equivalent to the system

$$\phi_x = \frac{1}{f^2}\psi_y, \qquad \phi_y = -\frac{1}{f^2}\psi_x.$$

In other words W is a solution of the main Vekua equation iff its corresponding pseudoanalytic function of the second kind is f^2-analytic. Thus, we obtain the following connection between the stationary Schrödinger equation and the system defining *p*-analytic functions.

Theorem 48. *Let f be a positive solution of the equation*

$$-\Delta g + \nu g = 0 \tag{3.37}$$

where ν is a real-valued function and let W_1 be another real-valued solution of this equation. Then the function $\Phi = W_1/f + ifW_2$, where W_2 is defined by (3.26) is an f^2-analytic function, and vice versa, let Φ be an f^2-analytic function then the function $W_1 = f \operatorname{Re}\Phi$ is a solution of (3.37).

The following relation between solutions of the conductivity equation and *p*-analytic functions is valid also.

Theorem 49. *Let f be a positive continuously differentiable function in a domain Ω and let U be a real-valued solution of the equation*

$$\operatorname{div}(f^2\nabla U) = 0 \qquad in \ \Omega. \tag{3.38}$$

Then the function $\Phi = U + iV$ is f^2-analytic in Ω, where V is defined by (3.29), and vice versa, let Φ be f^2-analytic in Ω then $U = \operatorname{Re}\Phi$ is a solution of (3.38).

Thus, solutions of the stationary Schrödinger equation and of the conductivity equation can be converted into *p*-analytic functions and vice versa. In some cases this relation leads to a simplification of a part of *p*-analytic function theory.

Example 50. One of the most important and intensely studied classes of *p*-analytic functions are x^k-analytic functions, where k is any integer number (see, e.g., [2], [27], [47], [59], [99], [100], [103], [126], [127]). Let us see what the form is of the corresponding Schrödinger equation. For this we should calculate the potential ν in (3.37) when $f = x^{k/2}$. It is easy to see that

$$\nu = \frac{k^2 - 2k}{4x^2}. \tag{3.39}$$

The Schrödinger equation with this potential has been well studied. Separation of variables leads us to the equation

$$X''(x) + \left(\beta^2 - \frac{4\alpha^2 - 1}{4x^2}\right) X(x) = 0, \tag{3.40}$$

where β^2 is the separation constant and $\alpha = (k-1)/2$. The function

$$X(x) = \sqrt{x} Z_\alpha(\beta x)$$

is a solution of (3.40) (see [51, 8.491]) where Z_α denotes any cylindric function of order α (Bessel functions of first or second kind). Thus the study of x^k-analytic functions reduces to the Schrödinger equation (3.1) with ν defined by (3.39) which in its turn, after having separated variables, reduces to a kind of Bessel equation (3.40).

Example 51. In the work [58] boundary value problems for *p*-analytic functions with $p = x/(x^2 + y^2)$ were studied. Considering

$$f = \sqrt{p} = \sqrt{\frac{x}{x^2 + y^2}}$$

we see that this function is a solution of the Schrödinger equation (3.1) with ν having the form

$$\nu = -\frac{1}{4x^2}.$$

That is, again we obtain a potential of the form (3.39) where $k = 1$ and, as was shown in the previous remark, the study of corresponding *p*-analytic functions in a sense reduces to the Bessel equation (3.40).

Chapter 4

Formal Powers

4.1 Definition

A generating sequence defines an infinite sequence of Vekua equations. If for a given (original) Vekua equation we know not only a corresponding generating pair but the whole generating sequence, that is a pair of exact and independent solutions for each of the Vekua equations from the infinite sequence of equations corresponding to the original one, we are able to construct an infinite system of solutions of the original Vekua equation as is shown in the next definition. Moreover, as we show in this chapter, under quite general conditions this infinite system of solutions is complete.

Definition 52. [13] The formal power $Z_m^{(0)}(a, z_0; z)$ with center at $z_0 \in \Omega$, coefficient a and exponent 0 is defined as the linear combination of the generators F_m, G_m with real constant coefficients λ, μ chosen so that $\lambda F_m(z_0) + \mu G_m(z_0) = a$. The formal powers with exponents $n = 1, 2, \ldots$ are defined by the recursion formula

$$Z_m^{(n)}(a, z_0; z) = n \int_{z_0}^{z} Z_{m+1}^{(n-1)}(a, z_0; \zeta) d_{(F_m, G_m)} \zeta. \tag{4.1}$$

This definition implies the following properties.

1. $Z_m^{(n)}(a, z_0; z)$ is an (F_m, G_m)-pseudoanalytic function of z.
2. If a' and a'' are real constants, then $Z_m^{(n)}(a' + ia'', z_0; z) = a' Z_m^{(n)}(1, z_0; z) + a'' Z_m^{(n)}(i, z_0; z)$.
3. The formal powers satisfy the differential relations

$$\frac{d_{(F_m, G_m)} Z_m^{(n)}(a, z_0; z)}{dz} = n Z_{m+1}^{(n-1)}(a, z_0; z).$$

4. The asymptotic formulas

$$Z_m^{(n)}(a, z_0; z) \sim a(z - z_0)^n, \quad z \to z_0 \tag{4.2}$$

hold.

Due to Property 2 of formal powers we have that $Z^{(n)}(a, z_0; z)$ for any coefficient a can be expressed through $Z^{(n)}(1, z_0; z)$ and $Z^{(n)}(i, z_0; z)$. Thus for any n it is sufficient to calculate only these two formal powers.

The question on the completeness of. the system of formal powers in the space of all solutions of a certain Vekua equation will be considered in detail in subsequent sections. Here we give an example of constructed formal powers.

Example 53. Consider the Yukawa equation

$$(-\Delta + c^2)u = 0 \tag{4.3}$$

with c being a real constant. Take the following particular solution of (4.3) $f = e^{cy}$. The corresponding main Vekua equation has the form

$$W_{\bar{z}} = \frac{ic}{2}\overline{W}. \tag{4.4}$$

Note that the generating pair $(F, G) = (f(y), i/f(y))$ is embedded into a periodic generating sequence with the period 1. That is in this case the generating sequence can be chosen in such a way that $(F_m, G_m) = (F, G)$ for any integer m. Let us construct the first few corresponding formal powers with center at the origin. We have

$$Z^{(0)}(1, 0; z) = e^{cy}, \qquad Z^{(0)}(i, 0; z) = ie^{-cy},$$

$$Z^{(1)}(1, 0; z) = xe^{cy} + \frac{i\sinh(cy)}{c}, \qquad Z^{(1)}(i, 0; z) = -\frac{\sinh(cy)}{c} + ixe^{-cy},$$

$$Z^{(2)}(1, 0; z) = \left(x^2 - \frac{y}{c}\right)e^{cy} + \frac{\sinh(cy)}{c^2} + \frac{2ix\sinh(cy)}{c},$$

$$Z^{(2)}(i, 0; z) = -\frac{2x\sinh(cy)}{c} + i\left(\left(x^2 + \frac{y}{c}\right)e^{-cy} - \frac{\sinh(cy)}{c^2}\right), \dots.$$

It is easy to check that each of these functions is indeed a solution of (4.4) and satisfies the property (4.2). Consider for instance $Z^{(1)}(1, 0; z)$ and use the Taylor series expansions of the elementary functions involved. We obtain

$$Z^{(1)}(1, 0; z) = x\left(1 + cy + \frac{(cy)^2}{2!} + \cdots\right) + \frac{i}{c}\left(cy + \frac{(cy)^3}{3!} + \cdots\right)$$

$$= x + iy + x\left(cy + \frac{(cy)^2}{2!} + \cdots\right) + \frac{i}{c}\left(\frac{(cy)^3}{3!} + \frac{(cy)^5}{5!} + \cdots\right)$$

$$\sim z, \quad \text{when } z \to 0.$$

Now taking real parts of the formal powers we obtain a complete system of solutions of the Yukawa equation:

$$u_1(x, y) = e^{cy}, \qquad u_2(x, y) = xe^{cy}, \qquad u_3(x, y) = -\frac{\sinh(cy)}{c},$$

$$u_4(x,y) = \left(x^2 - \frac{y}{c}\right)e^{cy} + \frac{\sinh(cy)}{c^2}, \qquad u_5(x,y) = -\frac{2x\sinh(cy)}{c}, \ldots$$

Formal powers of higher order can be constructed explicitly using a computer algebra system. For this particular example (studied together with Maria Rosalía Tenorio) Matlab 6.5 allowed us to obtain analytic expressions for the formal powers up to order ten, which gave us the first twenty-one functions $u_1, \ldots, u_{21}$.

The generating pair considered in this example $(F, G) = (f(y), i/f(y))$ belongs to a more general class of generating pairs for which L. Bers found elegant explicit formulas for corresponding formal powers which we give in the next section.

4.2 An important special case

Here following [13] we give explicit formulas for the formal powers in the case when the generating pair has the form

$$F(x,y) = \frac{\sigma(x)}{\tau(y)} \quad \text{and} \quad G(x,y) = \frac{i\tau(y)}{\sigma(x)}$$

where σ and τ are real-valued functions of their corresponding variables. For simplicity we assume that $z_0 = 0$ and $F(0) = 1$. In this case the formal powers are constructed in an elegant manner as follows. First, denote

$$X^{(0)}(x) = \widetilde{X}^{(0)}(x) = Y^{(0)}(y) = \widetilde{Y}^{(0)}(y) = 1$$

and for $n = 1, 2, \ldots$ denote

$$X^{(n)}(x) = \begin{cases} n\int\limits_0^x X^{(n-1)}(\xi)\frac{d\xi}{\sigma^2(\xi)} & \text{for an odd } n, \\ n\int\limits_0^x X^{(n-1)}(\xi)\sigma^2(\xi)d\xi & \text{for an even } n, \end{cases}$$

$$\widetilde{X}^{(n)}(x) = \begin{cases} n\int\limits_0^x \widetilde{X}^{(n-1)}(\xi)\sigma^2(\xi)d\xi & \text{for an odd } n, \\ n\int\limits_0^x \widetilde{X}^{(n-1)}(\xi)\frac{d\xi}{\sigma^2(\xi)} & \text{for an even } n, \end{cases}$$

$$Y^{(n)}(y) = \begin{cases} n\int\limits_0^y Y^{(n-1)}(\eta)\frac{d\eta}{\tau^2(\eta)} & \text{for an odd } n, \\ n\int\limits_0^y Y^{(n-1)}(\eta)\tau^2(\eta)d\eta & \text{for an even } n, \end{cases}$$

$$\widetilde{Y}^{(n)}(y) = \begin{cases} n\int\limits_0^y \widetilde{Y}^{(n-1)}(\eta)\tau^2(\eta)d\eta & \text{for an odd } n, \\ n\int\limits_0^y \widetilde{Y}^{(n-1)}(\eta)\frac{d\eta}{\tau^2(\eta)} & \text{for an even } n. \end{cases}$$

Then for $a = a' + ia''$ we have

$$Z^{(n)}(a,0,z) = \frac{\sigma(x)}{\tau(y)} \operatorname{Re} {}_*Z^{(n)}(a,0,z) + \frac{i\tau(y)}{\sigma(x)} \operatorname{Im} {}_*Z^{(n)}(a,0,z)$$

where

$$_*Z^{(n)}(a,0,z) = a' \sum_{j=0}^{n} \binom{n}{j} X^{(n-j)} i^j Y^{(j)} \tag{4.5}$$

$$+ ia'' \sum_{j=0}^{n} \binom{n}{j} \widetilde{X}^{(n-j)} i^j \widetilde{Y}^{(j)} \quad \text{for an odd } n$$

and

$$_*Z^{(n)}(a,0,z) = a' \sum_{j=0}^{n} \binom{n}{j} \widetilde{X}^{(n-j)} i^j Y^{(j)} \tag{4.6}$$

$$+ ia'' \sum_{j=0}^{n} \binom{n}{j} X^{(n-j)} i^j \widetilde{Y}^{(j)} \quad \text{for an even } n.$$

In Part II we use this elegant and algorithmically simple construction due to Bers for obtaining a representation for solutions of Sturm-Liouville equations and in Section 4.8 we present a recent result which allows us to construct a generating sequence and consequently the corresponding formal powers in a much more general situation.

4.3   Similarity principle

We will need the following definition.

Definition 54. A function a is said to satisfy a Hölder condition (or to be Hölder continuous) in a domain Ω if there exist such positive constants C and ε, $0 < \varepsilon \leq 1$ that the inequality

$$|a(z_1) - a(z_2)| \leq C |z_1 - z_2|^\varepsilon$$

holds for any two points z_1 and z_2 of the domain Ω.

For more information and examples regarding Hölder continuous functions we refer, for example, to [46].

We will consider the Vekua equation

$$w_{\overline{z}} = aw + b\overline{w} \tag{4.7}$$

with coefficients a and b satisfying a Hölder condition and vanishing identically outside a large disk. These assumptions are only made for the sake of simplicity and can be substantially weakened (see [13], [14], [120]). Here we follow the exposition from [30] and [32].

Consider the integral

$$q(z) = T\rho(z) = -\frac{1}{\pi}\int_\Omega \frac{\rho(\zeta)d\xi d\eta}{\zeta - z} \tag{4.8}$$

where Ω is a bounded, simply connected domain, $\zeta = \xi + i\eta$ and ρ is a complex-valued Hölder continuous function defined everywhere and vanishing outside a disk of radius R.

Theorem 55 (see, e.g., [32]). *Let the continuous complex-valued function ρ vanish identically outside a disk $|z| < R$ and everywhere satisfy the inequality $|\rho| \leq M$. Then*

$$|q(z)| \leq \frac{KM}{1 + |z|^\varepsilon}, \qquad |q(z_1) - q(z_2)| \leq KM\,|z_1 - z_2|^\varepsilon$$

for any ε such that $0 < \varepsilon < 1$, where q is defined by (4.8) and the constant K depends only on ε and R.

If additionally, the function ρ is Hölder continuous in Ω, then q possesses Hölder continuous partial derivatives and satisfies the equation

$$q_{\overline{z}} = \rho \quad in\ \Omega.$$

Proof. For the detailed proof of this theorem we refer to [30]. $\square$

Theorem 56. *A continuous bounded function w defined in Ω is a solution of the Vekua equation (4.7) if and only if the function*

$$\Psi = w - T[aw + b\overline{w}] \tag{4.9}$$

is analytic in Ω.

Proof. Note that $T[aw+b\overline{w}]$ is Hölder continuous and hence w is Hölder continuous if and only if Ψ is. Moreover, w is continuously differentiable if and only if Ψ is. Hence from Theorem 55 we have that, if either Ψ is analytic or w is pseudoanalytic, then

$$\Psi_{\overline{z}} = w_{\overline{z}} - aw - b\overline{w} = 0 \quad in\ \Omega. \qquad \square$$

Theorem 57 (Removable singularity theorem). *Let w be pseudoanalytic and bounded for $0 < |z - z_0| < r$ for some number $r > 0$. Then w can be defined at z_0 in such a way that it is pseudoanalytic in the whole disk $|z - z_0| < r$.*

Proof. The proof follows from the preceding theorem, the removable singularity theorem in analytic function theory and the fact that the integral in (4.9) is unchanged if a point is removed from its region of integration. $\square$

Theorem 58 (Similarity principle). *Let w be pseudoanalytic in a domain Ω. Then there exists an analytic function Φ defined in Ω and a Hölder continuous function s defined in $\overline{\Omega}$ such that*

$$w = \Phi e^s, \tag{4.10}$$

and vice versa, let Φ be an analytic function defined in a domain Ω. Then there exists a function s, Hölder continuous in $\overline{\Omega}$ such that (4.10) is pseudoanalytic in Ω. The function s can be constructed in such a way that $s(z_0) = 0$ at any fixed point $z_0 \in \Omega$.

Proof. The idea of the proof is simple. Given a pseudoanalytic function w, consider the functions

$$s = T\left[a + b\frac{\overline{w}}{w}\right]$$

and

$$\Phi = e^{-s}w. \tag{4.11}$$

Apply $\partial_{\overline{z}}$ to Φ:

$$\partial_{\overline{z}}\Phi = -\partial_{\overline{z}}s \cdot e^{-s}w + e^{-s}w_{\overline{z}}$$

$$= -\left(a + b\frac{\overline{w}}{w}\right)e^{-s}w + e^{-s}(aw + b\overline{w}) = 0.$$

Thus, Φ is an analytic function in Ω. In order to make this reasoning rigorous we should prove that s is differentiable and consider the case when w has zeros in Ω.

Assume that w is not identically zero (otherwise there is nothing to prove) and let Ω_0 be the open subset of Ω in which $w(z) \neq 0$, $z \in \Omega_0$. By Theorem 55, s is continuously differentiable in Ω_0 and therefore application of the operator $\partial_{\overline{z}}$ to Φ defined by (4.11) makes sense and Φ is analytic in Ω_0. The analyticity of Φ in $\Omega \backslash \Omega_0$ is proved with the aid of Theorem 57. We omit this part of the proof, referring the reader to [30, p. 142] or [32, Chapter 4].

The second part of the theorem is based mainly on the same ideas and can also be found in [30, p. 142] or [32, Chapter 4].   □

From the similarity principle a variety of function theoretic results for pseudoanalytic functions can be deduced.

Corollary 59 (Carleman's theorem). *A pseudoanalytic function which does not vanish identically has only isolated zeros.*

Proof. This follows immediately from Theorem 58 and the corresponding classical results from analytic function theory.   □

Corollary 60 (Liouville's theorem). *A bounded pseudoanalytic function defined in the entire plane and vanishing at a fixed point z_0 of the plane (z_0 may be infinity) is identically zero.*

Proof. By Theorem 58, $w = \Phi e^s$ where Φ is an entire function and s is bounded (due to the fact that a and b have compact support). Thus we have that Φ is bounded and hence, by Liouville's theorem in analytic function theory, is a constant. Since $w(z_0) = 0$, this constant must be zero, and the corollary follows.   □

Corollary 61 (Maximum modulus theorem). *Let w be a pseudoanalytic function defined in Ω and continuous in $\overline{\Omega}$. Then there exists a positive constant $M \geq 1$ depending only on a and b such that, for z in Ω,*

$$|w(z)| \leq M \max_{t \in \partial\Omega} |w(t)|.$$

Proof. This follows from Theorem 58 and the maximum modulus principle for analytic functions. $\square$

4.4 Taylor series in formal powers

Assume now that

$$W(z) = \sum_{n=0}^{\infty} Z^{(n)}(a_n, z_0; z) \tag{4.12}$$

where the absence of the subindex m means that all the formal powers correspond to the same generating pair (F, G) and the series converges uniformly in some neighborhood of z_0. Based on the similarity principle it can be shown [15] that the uniform limit of pseudoanalytic functions is pseudoanalytic, and that a uniformly convergent series of (F, G)-pseudoanalytic functions can be (F, G)-differentiated term by term. Hence the function W in (4.12) is (F, G)-pseudoanalytic and its rth derivative admits the expansion

$$W^{[r]}(z) = \sum_{n=r}^{\infty} n(n-1)\cdots(n-r+1)Z_r^{(n-r)}(a_n, z_0; z).$$

From this the Taylor formulas for the coefficients are obtained,

$$a_n = \frac{W^{[n]}(z_0)}{n!}. \tag{4.13}$$

Definition 62. Let $W(z)$ be a given (F, G)-pseudoanalytic function defined for small values of $|z - z_0|$. The series

$$\sum_{n=0}^{\infty} Z^{(n)}(a_n, z_0; z) \tag{4.14}$$

with the coefficients given by (4.13) is called the Taylor series of W at z_0, formed with formal powers.

The Taylor series always represents the function asymptotically:

$$W(z) - \sum_{n=0}^{N} Z^{(n)}(a_n, z_0; z) = O\left(|z - z_0|^{N+1}\right), \quad z \to z_0, \tag{4.15}$$

for all N. This implies (since a pseudoanalytic function can not have a zero of arbitrarily high order without vanishing identically) that the sequence of derivatives $\{W^{[n]}(z_0)\}$ determines the function W uniquely.

If the series (4.14) converges uniformly in a neighborhood of z_0, it converges to the function W.

The statements given in this section were obtained by L. Bers [13], [15] and S. Agmon and L. Bers [1].

Theorem 63 ([13]). *The formal Taylor expansion* (4.14) *of a pseudoanalytic function in formal powers defined by a periodic generating sequence converges in some neighborhood of the center.*

This theorem means only a local completeness of the system of formal powers. The following definition due to L. Bers describes the case when corresponding formal powers represent a globally complete system of solutions of a Vekua equation much as in the case of usual powers of the variable z and the Cauchy-Riemann equation.

Definition 64. [13] A generating pair (F, G) is called complete if these functions are defined and satisfy the Hölder condition for all finite values of z, the limits $F(\infty)$, $G(\infty)$ exist, $\mathrm{Im}(\overline{F(\infty)}G(\infty)) > 0$, and the functions $F(1/z)$, $G(1/z)$ also satisfy the Hölder condition. A complete generating pair is called normalized if $F(\infty) = 1$, $G(\infty) = i$.

A generating pair equivalent to a complete one is complete, and every complete generating pair is equivalent to a uniquely determined normalized pair. The adjoint of a complete (normalized) generating pair is complete (normalized).

From now on we assume that (F, G) is a complete normalized generating pair. Then much more can be said on the series of corresponding formal powers. We limit ourselves to the following completeness results (the expansion theorem and Runge's approximation theorem for pseudoanalytic functions).

Following [13] we shall say that a sequence of functions W_n converges normally in a domain Ω if it converges uniformly on every bounded closed subdomain of Ω.

Theorem 65. *Let W be an (F, G)-pseudoanalytic function defined for $|z - z_0| < R$. Then it admits a unique expansion of the form $W(z) = \sum_{n=0}^{\infty} Z^{(n)}(a_n, z_0; z)$ which converges normally for $|z - z_0| < \theta R$, where θ is a positive constant depending on the generating sequence.*

The first version of this theorem was proved in [1]. We follow here [15].

Remark 66. Necessary and sufficient conditions for the relation $\theta = 1$ are, unfortunately, not known. However, in [15] the following sufficient conditions for the case when the generators (F, G) possess partial derivatives are given. One such condition reads:

$$|F_{\bar{z}}(z)| + |G_{\bar{z}}(z)| \leq \frac{\mathrm{Const}}{1 + |z|^{1+\varepsilon}}$$

for some $\varepsilon > 0$. Another condition is

$$\int\int_{|z|<\infty} \left(|F_{\overline{z}}|^{2-\varepsilon} + |F_{\overline{z}}|^{2+\varepsilon} + |G_{\overline{z}}|^{2-\varepsilon} + |G_{\overline{z}}|^{2+\varepsilon} \right) dxdy < \infty$$

for some $0 < \varepsilon < 1$.

4.5 The Runge theorem

The following theorem proved by L. Bers is a generalization of the well-known Runge theorem from complex analysis.

Theorem 67 ([15]). *A pseudoanalytic function defined in a simply connected domain can be expanded into a normally convergent series of formal polynomials (linear combinations of formal powers with positive exponents).*

Remark 68. This theorem admits a direct generalization onto the case of a multiply-connected domain (see [15]).

In subsequent works [54], [91], [44] and others, deep results on interpolation and on the degree of approximation by pseudopolynomials were obtained. For example,

Theorem 69 ([91]). *Let W be a pseudoanalytic function in a domain Ω bounded by a Jordan curve and satisfy the Hölder condition on $\partial\Omega$ with the exponent α ($0 < \alpha \le 1$). Then for any $\varepsilon > 0$ and any natural n there exists a pseudopolynomial of order n satisfying the inequality*

$$|W(z) - P_n(z)| \le \frac{\text{Const}}{n^{\alpha-\varepsilon}} \qquad \text{for any } z \in \overline{\Omega}$$

where the constant does not depend on n, but only on ε.

4.6 Complete systems of solutions for second-order equations

In what follows let us suppose that Ω is a bounded simply connected domain and the positive function $f \in C^1(\overline{\Omega})$ is defined in a somewhat bigger domain Ω_ε with a sufficiently smooth boundary. Then we change the function f for $z \in \Omega_\varepsilon \backslash \Omega$ and continue it over the whole plane in such a way that $f \equiv 1$ for large $|z|$ (see [15]) and $f(z)$ together with $f(1/z)$ satisfy the Hölder condition for all finite values of z. In this way the generating pair

$$(F, G) = (f, i/f) \tag{4.16}$$

becomes complete and normalized.

Then the following statements are direct corollaries of the relations established in Section 3.2 between pseudoanalytic functions (solutions of (3.15)) and solutions of second-order elliptic equations, and of the convergence theorems from sections 4.4 and 4.5.

Definition 70. [71] Let $u(z)$ be a given solution of the equation (3.10) defined for small values of $|z - z_0|$, and let $W(z)$ be a solution of (3.15) constructed according to Corollary 39, such that $\operatorname{Re} W = p^{1/2} u$. The series

$$p^{-1/2}(z) \sum_{n=0}^{\infty} \operatorname{Re} Z^{(n)}(a_n, z_0; z)$$

with the coefficients given by (4.13) is called the Taylor series of u at z_0, formed with formal powers.

Theorem 71 ([71]). *Let $u(z)$ be a solution of (3.10) defined for $|z - z_0| < R$. Then it admits a unique expansion of the form*

$$u(z) = p^{-1/2}(z) \sum_{n=0}^{\infty} \operatorname{Re} Z^{(n)}(a_n, z_0; z)$$

which converges normally for $|z - z_0| < R$.

Proof. This is a direct consequence of Theorem 65 and Remark 66. Both necessary conditions in Remark 66 are fulfilled for the generating pair (4.16). $\square$

Theorem 72 ([71]). *An arbitrary solution of (3.10) defined in a bounded simply connected domain Ω where there exists a positive particular solution $u_0 \in C^1(\overline{\Omega})$ can be expanded into a normally convergent series of formal polynomials multiplied by $p^{-1/2}$.*

Proof. This is a direct corollary of Theorem 67. $\square$

More precisely the last theorem has the following meaning. Due to Property 2 of formal powers we have that, for any Taylor coefficient a, the formal power $Z^{(n)}(a, z_0; z)$ can be expressed through $Z^{(n)}(1, z_0; z)$ and $Z^{(n)}(i, z_0; z)$. Then due to Theorem 67 any solution W of (3.15) can be expanded into a normally convergent series of linear combinations of $Z^{(n)}(1, z_0; z)$ and $Z^{(n)}(i, z_0; z)$. Consequently, any solution of (3.10) can be expanded into a normally convergent series of linear combinations of real parts of $Z^{(n)}(1, z_0; z)$ and $Z^{(n)}(i, z_0; z)$ multiplied by $p^{-1/2}$.

Obviously, for solutions of (3.10) the results on the interpolation and on the degree of approximation like, e.g., Theorem 69 are also valid.

Let us stress that Theorem 72 gives us the following result. The functions

$$\left\{ p^{-1/2}(z) \operatorname{Re} Z^{(n)}(1, z_0; z), \quad p^{-1/2}(z) \operatorname{Re} Z^{(n)}(i, z_0; z) \right\}_{n=0}^{\infty} \qquad (4.17)$$

represent a complete system of solutions of (3.10) in the sense that any solution of (3.10) can be represented by a normally convergent series formed by functions (4.17) in any bounded simply connected domain Ω where a positive solution of (3.10) exists. Moreover, as we show in Section 4.8, in many practically interesting situations these functions can be constructed explicitly.

4.7 A remark on orthogonal coordinate systems in a plane

Orthogonal coordinate systems in a plane are obtained (see [87]) from Cartesian coordinates x, y by means of the relation

$$u + iv = \Phi(x + iy)$$

where Φ is an arbitrary analytic function. Quite often a transition to more general coordinates is useful,

$$\xi = \xi(u), \quad \eta = \eta(v).$$

ξ and η preserve the property of orthogonality. Some examples taken from [87] illustrate the point.

Example 73 (Polar coordinates).

$$u + iv = \ln(x + iy),$$

$$u = \ln\sqrt{x^2 + y^2}, \quad v = \arctan\frac{y}{x}. \tag{4.18}$$

Usually the following new coordinates are introduced:

$$r = e^u = \sqrt{x^2 + y^2}, \quad \varphi = v = \arctan\frac{y}{x}.$$

Example 74 (Parabolic coordinates).

$$\frac{u + iv}{\sqrt{2}} = \sqrt{x + iy},$$

$$u = \sqrt{r + x}, \quad v = \sqrt{r - x}.$$

More frequently the parabolic coordinates are introduced as

$$\xi = u^2, \quad \eta = v^2.$$

Example 75 (Elliptic coordinates).

$$u + iv = \arcsin\frac{x + iy}{\alpha},$$

$$\sin u = \frac{s_1 - s_2}{2\alpha}, \quad \cosh v = \frac{s_1 + s_2}{2\alpha}$$

where $s_1 = \sqrt{(x+\alpha)^2 + y^2}$, $s_2 = \sqrt{(x-\alpha)^2 + y^2}$. The substitution

$$\xi = \sin u, \quad \eta = \cosh v$$

is frequently used.

Example 76 (Bipolar coordinates).

$$u + iv = \ln \frac{\alpha + x + iy}{\alpha - x - iy},$$

$$\tanh u = \frac{2\alpha x}{\alpha^2 + x^2 + y^2}, \quad \tan v = \frac{2\alpha y}{\alpha^2 - x^2 - y^2}.$$

The following substitution is frequently used:

$$\xi = e^{-u}, \quad \eta = \pi - v.$$

4.8   Explicit construction of a generating sequence

We suppose that the function f in the main Vekua equation (3.15) has the form

$$f = U(u)V(v) \tag{4.19}$$

where u and v represent an orthogonal coordinate system and, according to the explanation in the previous section, we assume that $\Phi = u + iv$ is an analytic function of the variable $z = x + iy$. U and V are arbitrary differentiable nonvanishing real-valued functions.

The present section is based on the results from [73] and structured in the following way. First we explain how one can naturally arrive at the form of generating sequences in the main result of the section, Theorem 77 and then we give a rigorous proof of this result.

The first step in the construction of a generating sequence for the main Vekua equation (3.15) is the construction of a generating pair for the equation (3.4) which as was shown in Section 3.4 is a successor of the main Vekua equation. For this one of the possibilities consists in constructing another pair of solutions of (3.15). Then their (F, G)-derivatives will give us solutions of (3.4). Let us consider equation (3.15) in its equivalent form (3.22):

$$\varphi_{\bar{z}} + \frac{i}{f^2}\psi_{\bar{z}} = 0 \tag{4.20}$$

and look for a solution in the form

$$\varphi = \varphi(u), \quad \psi = \psi(v).$$

Then we have the equation

$$\varphi'(u)u_{\bar{z}} + \frac{i}{f^2}\psi'(v)v_{\bar{z}} = 0.$$

Taking into account that $\Phi = u + iv$ is an analytic function, that is $u_{\bar{z}} + iv_{\bar{z}} = 0$, we observe that (4.20) is fulfilled if $\varphi'(u) = \psi'(v)/f^2$. Now using (4.19) we obtain that $U^2(u)V^2(v) = \psi'(v)/\varphi'(u)$ and hence

$$\varphi(u) = \int \frac{du}{U^2(u)} \quad \text{and} \quad \psi(v) = \int V^2(v)dv.$$

The corresponding solution W_1 of (3.15) has the form

$$W_1 = \int \frac{du}{U^2(u)}U(u)V(v) + \int V^2(v)dv\frac{i}{U(u)V(v)}.$$

Its (F,G)-derivative is obtained according to (2.10) as

$$\dot{W}_1 = \frac{V}{U}u_z + i\frac{V}{U}v_z = \frac{V}{U}\Phi_z.$$

By analogy, we can look for a solution of (4.20) in the form

$$\varphi = \varphi(v), \qquad \psi = \psi(u).$$

Then we have the equation

$$\varphi'(v)v_{\bar{z}} + \frac{i}{f^2}\psi'(u)u_{\bar{z}} = 0.$$

This equation is fulfilled if $\varphi'(v) = -1/V^2(v)$ and $\psi'(u) - U^2(u)$. Consequently,

$$\varphi(v) = -\int \frac{dv}{V^2(v)} \quad \text{and} \quad \psi(u) = \int U^2(u)du.$$

The corresponding solution W_2 of (3.15) has the form

$$W_2 = -\int \frac{dv}{V^2(v)}U(u)V(v) + \int U^2(u)du\frac{i}{U(u)V(v)}$$

with its corresponding (F,G)-derivative

$$\dot{W}_2 = -\frac{U}{V}v_z + i\frac{U}{V}u_z = i\frac{U}{V}\Phi_z.$$

Write $F_1 = \dot{W}_1$ and $G_1 = \dot{W}_2$. It is easy to see that the pair F_1, G_1 fulfils (2.1) and by construction satisfies equation (3.4). Thus, (F_1, G_1) is a successor of (F, G) defined by (3.21).

The following step is to construct the generating pair (F_2, G_2). For this we should find another pair of solutions of (3.4) and apply the (F_1, G_1)-derivative to them. Consider equation (3.4) written in the form

$$\varphi_{\bar{z}} F_1 + \psi_{\bar{z}} G_1 = 0$$

which in our case can be represented as

$$\varphi_{\bar{z}} \frac{V}{U} + \psi_{\bar{z}} i \frac{U}{V} = 0. \tag{4.21}$$

Again, let us look for a solution in the form

$$\varphi = \varphi(u), \qquad \psi = \psi(v).$$

Then equation (4.21) is satisfied if $\varphi'(u) = \left(\frac{U(u)}{V(v)}\right)^2 \psi'(v)$ from which we obtain

$$\varphi(u) = \int U^2(u) du \quad \text{and} \quad \psi(v) = \int V^2(v) dv.$$

Then the corresponding solution of (3.4) has the form

$$w_1 = \int U^2(u) du \Phi_z \frac{V(v)}{U(u)} + \int V^2(v) dv \Phi_z \frac{iU(u)}{V(v)}.$$

Its (F_1, G_1)-derivative is obtained as

$$\dot{w}_1 = \Phi_z UV u_z + i\Phi_z UV v_z = UV (\Phi_z)^2.$$

Analogously, looking for a solution of (4.21) in the form

$$\varphi = \varphi(v), \qquad \psi = \psi(u)$$

we obtain that

$$\varphi(v) = -\int \frac{dv}{V^2(v)} \quad \text{and} \quad \psi(u) = \int \frac{du}{U^2(u)}.$$

The corresponding solution of (3.4) has the form

$$w_2 = -\int \frac{dv}{V^2(v)} \Phi_z \frac{V(v)}{U(u)} + \int \frac{du}{U^2(u)} \Phi_z \frac{iU(u)}{V(v)}.$$

Its (F_1, G_1)-derivative is obtained as

$$\dot{w}_2 = -\frac{\Phi_z v_z}{UV} + \frac{\Phi_z i u_z}{UV} = \frac{i}{UV} (\Phi_z)^2.$$

We have that the pair $F_2 = \dot{w}_1$ and $G_2 = \dot{w}_2$ is a successor of (F_1, G_1). Observe that

$$(F_1, G_1) = \left(\frac{\Phi_z}{U^2} F, \quad U^2 \Phi_z G \right)$$

and

$$(F_2, G_2) = \left((\Phi_z)^2 F, \quad (\Phi_z)^2 G \right).$$

From these formulae it is easy to guess a general form of the corresponding generating sequence which we give in the following statement.

Theorem 77 ([73]). *Let* $F = U(u)V(v)$ *and* $G = \frac{i}{U(u)V(v)}$ *where* U *and* V *are arbitrary differentiable nonvanishing real-valued functions,* $\Phi = u + iv$ *is an analytic function of the variable* $z = x + iy$ *in* Ω *such that* Φ_z *is bounded and has no zeros in* Ω. *Then the generating pair* (F, G) *is embedded in the generating sequence* (F_m, G_m), $m = 0, \pm 1, \pm 2, \ldots$ *in* Ω *defined as*

$$F_m = (\Phi_z)^m F \quad and \quad G_m = (\Phi_z)^m G \quad for \ even \ m$$

and

$$F_m = \frac{(\Phi_z)^m}{U^2} F \quad and \quad G_m = (\Phi_z)^m U^2 G \quad for \ odd \ m.$$

Proof. First of all let us show that (F_m, G_m) is a generating pair for $m = \pm 1$, $\pm 2, \ldots$. Indeed we have

$$\text{Im}(\overline{F}_m G_m) = \text{Im}(|\Phi_z|^{2m} \overline{F}G) > 0.$$

Consider $a_{(F_m, G_m)}$. For both m being even or odd we obtain

$$a_{(F_m, G_m)} = |\Phi_z|^{2m} a_{(F,G)} \equiv 0.$$

We should verify the equality

$$b_{(F_m, G_m)} = -B_{(F_{m-1}, G_{m-1})}. \tag{4.22}$$

Consider first the case of an odd m. A direct calculation gives us

$$b_{(F_m, G_m)} = \left(\frac{\Phi_z}{\overline{\Phi}_z} \right)^m \left(b_{(F,G)} - 2u_{\overline{z}} \frac{U'}{U} \right)$$

and

$$B_{(F_{m-1}, G_{m-1})} = \left(\frac{\Phi_z}{\overline{\Phi}_z} \right)^{m-1} B_{(F,G)}.$$

Thus equality (4.22) is true iff $\frac{\Phi_z}{\overline{\Phi}_z} \left(b_{(F,G)} - 2u_{\overline{z}} \frac{U'}{U} \right)$ is equal to $-B_{(F,G)}$. It is easy to see that

$$B_{(F,G)} = u_z \left(\frac{U'}{U} - i \frac{V'}{V} \right).$$

Consider

$$\frac{\Phi_z}{\overline{\Phi_z}}\left(b_{(F,G)} - 2u_{\overline{z}}\frac{U'}{U}\right) = \frac{u_z + iv_z}{u_{\overline{z}} - iv_{\overline{z}}}\left(\frac{U'}{U}u_{\overline{z}} + \frac{V'}{V}v_{\overline{z}} - 2u_{\overline{z}}\frac{U'}{U}\right)$$

$$= \frac{u_z}{u_{\overline{z}}}\left(-\frac{U'}{U}u_{\overline{z}} + \frac{V'}{V}iu_{\overline{z}}\right) = u_z\left(-\frac{U'}{U} + i\frac{V'}{V}\right).$$

Thus, equality (4.22) is proved in the case of m being odd.

Now let m be even. Then

$$b_{(F_m,G_m)} = \left(\frac{\Phi_z}{\overline{\Phi_z}}\right)^m b_{(F,G)}$$

and

$$B_{(F_{m-1},G_{m-1})} = \left(\frac{\Phi_z}{\overline{\Phi_z}}\right)^{m-1}\left(B_{(F,G)} - 2\frac{U'}{U}u_z\right).$$

Equality (4.22) is valid iff the expression $\frac{\Phi_z}{\overline{\Phi_z}}b_{(F,G)}$ is equal to $\left(-B_{(F,G)} + 2\frac{U'}{U}u_z\right)$. We have

$$\frac{\Phi_z}{\overline{\Phi_z}}b_{(F,G)} = \frac{u_z + iv_z}{u_{\overline{z}} - iv_{\overline{z}}}\left(\frac{U'}{U}u_{\overline{z}} + \frac{V'}{V}v_{\overline{z}}\right) = \frac{u_z}{u_{\overline{z}}}\left(\frac{U'}{U}u_{\overline{z}} + \frac{V'}{V}iu_{\overline{z}}\right) = u_z\left(\frac{U'}{U} + i\frac{V'}{V}\right)$$

and from the other side

$$-B_{(F,G)} + 2\frac{U'}{U}u_z = -u_z\left(\frac{U'}{U} - i\frac{V'}{V}\right) + 2\frac{U'}{U}u_z = u_z\left(\frac{U'}{U} + i\frac{V'}{V}\right).$$

Thus, equality (4.22) is proved in all cases and the sequence (F_m, G_m), $m = 0, \pm1, \pm2, \ldots$ satisfies the conditions of Definition 12. Therefore it is a generating sequence. $\qquad\square$

Remark 78. This result obviously generalizes the explicit construction of a generating sequence in the case when $u = x$ and $v = y$ presented in Section 4.2.

The last theorem opens the way for explicit construction of formal powers corresponding to the main Vekua equation (3.15) in the case when f has the form (4.19) and hence for explicit construction of complete systems of solutions for corresponding second-order elliptic equations. We give more details on this as well as some examples in the next section.

4.9   Explicit construction of complete systems of solutions of second-order elliptic equations

In Section 4.6 we explained how complete systems of solutions of second-order elliptic equations can be constructed from systems of formal powers for a corresponding main Vekua equation (3.15). In the preceding section we established a result which allows us to make this construction possible in the case when f in (3.15) has the form (4.19). Then formal powers are constructed simply by Definition 52 using the generating sequence constructed in Theorem 77. The meaning of this result for the stationary Schrödinger equation and for the conductivity equation we discuss separately in the next two subsections.

4.9.1 Explicit construction of complete systems of solutions for a stationary Schrödinger equation

Consider the equation

$$-\Delta g + \nu g = 0 \qquad \text{in } \Omega \tag{4.23}$$

where ν is a real-valued function. In order to start the procedure of construction of a complete system of solutions of this equation we need a positive particular solution f having the form (4.19).

Example 79. Let ν in (4.23) depend on one Cartesian variable: $\nu = \nu(x)$. Suppose we are given a particular solution $f = f(x)$ of the ordinary differential equation

$$-\frac{d^2 f}{dx^2} + \nu f = 0. \tag{4.24}$$

This solution is sufficient for the application of our result from the preceding section for constructing the corresponding generating sequence and hence the system of formal powers for the main Vekua equation which in this particular case has the form

$$W_{\overline{z}} = \frac{f_x}{2f} \overline{W}.$$

Example 80. A number of works (see, e.g., [23] and [24]) are dedicated to construction (in our terms) of a particular solution for the Schrödinger equation with a radially symmetric potential. This solution f is precisely the only necessary ingredient in order to obtain a complete system of formal powers and hence of solutions of the Schrödinger equation. Here an important restriction is that the analytic function Φ_z corresponding to the polar coordinate system has the form $1/z$ and hence our procedure works in any domain not including the origin and where f is positive.

4.9.2 Complete systems of solutions for the conductivity equation

Consider the equation

$$\text{div}(f^2 \nabla g) = 0.$$

Usually in practical applications the function f is positive and equals 1 outside a certain disk. We suppose additionally that f is continuously differentiable on the whole plane. Under these conditions $F = f$ and $G = i/f$ is a complete normalized generating pair and all the convergence and Runge type results are valid. Using results of Section 4.8 one can construct explicitly a complete system of solutions of the conductivity equation when the function f has the form (4.19) in any domain where the analytic function Φ_z from Theorem 77 is bounded and has no zeros.

4.10   Numerical solution of boundary value problems

In the case when Theorem 77 is applicable, the approximate solution of boundary value problems for the corresponding second-order equation can be done with the aid of the complete system of its solutions obtained as described in the preceding section. The numerical solution of the problem is sought in the form of a finite linear combination of the found exact solutions and the coefficients in the linear combination are obtained from the boundary conditions.

In Example 53 a complete system of exact solutions for the Yukawa equation was presented. We used these functions for a numerical solution of the corresponding Dirichlet problem with very satisfactory results. For example, in the case when Ω is a unit disk with centre at the origin, $c = 1$ and u on the boundary is equal to e^x (this test exact solution gave us the worst precision because of its obvious "disparateness" from functions $u_1, u_2 \ldots$) the maximal error $\max_{z \in \Omega} |u(z) - \widetilde{u}(z)|$ where u is the exact solution and $\widetilde{u} = \sum_{n=1}^{21} a_n u_n$, the real constants a_n being found by the collocation method, was of order 10^{-7}. A very fast convergence of the method was observed. Although the numerical method based on the usage of explicitly or numerically constructed pseudoanalytic formal powers still needs a much more detailed analysis, these results show us that it is quite possible that in due time it can rank high among other numerical approaches. The use of complete systems of exact solutions based on pseudoanalytic formal powers has an important advantage of universality. The system does not depend on the choice of a domain whenever the function f continues to have no zeros. The numerical computation of formal powers and of corresponding exact solutions of the second-order equations can be done with a remarkable precision.

Notice that in [20] the following system of solutions of the Yukawa equation (4.3) was obtained from completely different reasonings,

$$u_0 = \cosh(cy), \quad u_1 = x \cosh(cy), \quad u_2 = x^2 \cosh(cy) - \frac{y}{c} \sinh(cy),$$

$$u_3 = x^3 \cosh(cy) - \frac{3xy}{c} \sinh(cy), \ldots .$$

The same system is easily obtained (our Matlab program gives symbolic representations of these functions up to the subindex 21) if instead of the particular solution $f = e^{cy}$ considered in Example 53 we choose $f = \cosh(cy)$.

In the case of the Helmholtz equation

$$(\Delta + c^2)u = 0,$$

L.R. Bragg and J.W. Dettman constructed the system

$$u_0 = \cos(cy), \quad u_1 = x\cos(cy), \quad u_2 = x^2\cos(cy) - \frac{y}{c}\sin(cy),$$

$$u_3 = x^3\cos(cy) - \frac{3xy}{c}\sin(cy), \ldots.$$

This system is also easily obtained as explained above choosing $f = \cos(cy)$.

Thus, the construction of the systems of solutions for the Yukawa and for the Helmholtz equation as well as the proof of their completeness in [20] are corollaries of a more general theory presented here.

Chapter 5

Cauchy's Integral Formula

In this chapter we give information on the Cauchy integral formula for pseudo-analytic and p-analytic functions and using the results from [72] show how the Cauchy kernel can be constructed explicitly for x^k-analytic functions.

As was mentioned in Section 3.6, one of the most important and intensely studied classes of p-analytic functions are x^k-analytic functions, where k is any integer number. In spite of its importance and numerous attempts to obtain it explicitly the corresponding Cauchy integral formula was constructed in two cases only: when $k = 0$ (the classical Cauchy integral formula with the Cauchy kernel $1/(z - z_0)$) and when $k = 1$. In this case the Cauchy kernel has a complicated form (see [2], [33], [103]) involving elliptic integrals. We do not present it here, its simplification is an important task.

Here we introduce a procedure which allows us to obtain explicitly the Cauchy integral formulas for x^k-analytic functions, for any integer k, based on the knowledge of Cauchy kernels in the two above-mentioned cases ($k = 0, 1$). More precisely, our approach gives us recursive formulas for consecutive construction of the Cauchy kernels corresponding to even k starting from the known Cauchy kernel for $k = 0$ and to odd k beginning with the Cauchy kernel for $k = 1$. The procedure is based on a close relation between p-analytic functions and pseudoanalytic functions satisfying the main Vekua equation as well as on the concept of a transplant operator introduced in Section 5.3.

5.1 Preliminary information on the Cauchy integral formula for pseudoanalytic functions

The subject of Cauchy integral formulas for classes of pseudoanalytic functions has been treated by many different mathematical schools and with implementation of different notations and techniques. Here we follow the definitions and notations introduced by L. Bers due to their complete structural resemblance with the classical results from analytic function theory. Consider the Vekua equation

$$\partial_{\bar{z}} W = aW + b\overline{W} \tag{5.1}$$

in a simply connected, bounded domain Ω where a and b are complex-valued functions satisfying the Hölder condition up to the boundary. L. Bers (see [13]) proved the existence of a solution w of (5.1) in $\Omega \setminus \{z_0\}$ which satisfies the relation

$$\lim_{z \to z_0} \frac{w(z)}{\alpha(z - z_0)^{-1}} = 1 \tag{5.2}$$

where α is any complex number. This function is denoted as

$$w(z) = Z^{(-1)}(\alpha, z_0, z).$$

Note that, while no other condition is imposed, this function is not unique. The following generalization of the Cauchy integral formula is valid.

Theorem 81 ([13]). *Let W be a solution of (5.1) defined in a domain Ω bounded by a simple closed continuously differentiable curve Γ and assume that W is continuous up to the boundary Γ. Then for any $z \in \Omega$ the following equality holds:*

$$W(z) = \frac{1}{2\pi} \int_{\Gamma} Z^{(-1)}(iW(\zeta)d\zeta, \zeta, z). \tag{5.3}$$

This integral should be understood in the following sense. If the parametric representation of the curve Γ has the form $\zeta(t)$, $0 \le t \le T$, then for any function χ defined on Γ we have

$$\int_{\Gamma} Z^{(-1)}(\chi(\zeta)d\zeta, \zeta, z) = \int_0^T Z^{(-1)}(\chi(\zeta(t))\zeta'(t), \zeta(t), z)dt.$$

Let us stress that in (5.3) any solution of (5.1) in $\overline{\Omega} \setminus \{z_0\}$ satisfying (5.2) is acceptable in the role of a Cauchy kernel as in the case of analytic functions for the Cauchy integral formula in a bounded domain Ω the kernel can be taken, e.g., as $1/(z - z_0) + w(z)$, where w is any function analytic in $\overline{\Omega}$.

The Vekua equations considered in this chapter possess coefficients having singularities and the natural (physically meaningful) domain of interest Ω does not coincide with the whole complex plane.

Formal powers, and in particular the first negative formal power $Z^{(-1)}$, enjoy the following property. If α' and α'' are real constants then

$$Z^{(-1)}(\alpha' + i\alpha'', z_0, z) = \alpha' Z^{(-1)}(1, z_0, z) + \alpha'' Z^{(-1)}(i, z_0, z). \tag{5.4}$$

Thus, in order to be able to write down $Z^{(-1)}(\alpha, z_0, z)$ for any complex coefficient α, which is required for the Cauchy integral formula, we need to construct two Cauchy kernels:

$$Z^{(-1)}(1, z_0, z) \qquad \text{and} \qquad Z^{(-1)}(i, z_0, z).$$

We finish this section by noticing that the Cauchy integral from (5.3) possesses many important properties similar to those of a usual Cauchy integral including Plemelj-Sokhotski formulas and others, which are indispensable for solving corresponding boundary value problems.

5.2 Relation between the main Vekua equation and the system describing p-analytic functions

Let us consider the case when $a \equiv 0$ and $b = f_{\bar{z}}/f$ where f is a continuously differentiable real-valued positive function defined in Ω, that is the main Vekua equation. We write

$$\mathcal{V} := \partial_{\bar{z}} - \frac{f_{\bar{z}}}{f}C$$

where C is the operator of complex conjugation. We suppose that $p = f^2$ and introduce the operator

$$\Pi := f\partial_{\bar{z}}P^+ + \frac{1}{f}\partial_{\bar{z}}P^-,$$

where $P^{\pm} := \frac{1}{2}(I \pm C)$ and I is the identity operator. We have that the equation

$$\Pi\omega = 0 \tag{5.5}$$

is equivalent to the system

$$\varphi_x = \frac{1}{f^2}\psi_y, \qquad \varphi_y = -\frac{1}{f^2}\psi_x \tag{5.6}$$

where $\varphi = \operatorname{Re}\omega$ and $\psi = \operatorname{Im}\omega$.

We write

$$B := fP^+ + \frac{1}{f}P^-.$$

Then it is easy to see that

$$B^{-1} = \frac{1}{f}P^+ + fP^-.$$

Proposition 82 ([72]).

$$\mathcal{V}B = \Pi.$$

Proof. Consider

$$\left(\partial_{\bar{z}} - \frac{f_{\bar{z}}}{f}C\right)\left(fP^+ + \frac{1}{f}P^-\right)$$

$$= f\partial_{\bar{z}}P^+ + f_{\bar{z}}P^+ + \frac{1}{f}\partial_{\bar{z}}P^- - \frac{f_{\bar{z}}}{f^2}P^- - f_{\bar{z}}P^+ + \frac{f_{\bar{z}}}{f^2}P^- = f\partial_{\bar{z}}P^+ + \frac{1}{f}\partial_{\bar{z}}P^- = \Pi.$$

Here we used the fact that $CP^+ = P^+$ and $CP^- = -P^-$. $\qquad\square$

This proposition gives us a simple relation between the Vekua operator $\mathcal{V}$ and the operator Π corresponding to the system (5.6) describing p-analytic functions (with $p = f^2$) and in particular means that, if a function W satisfies the main Vekua equation

$$\left(\partial_{\bar{z}} - \frac{f_{\bar{z}}}{f}C\right)W = 0, \tag{5.7}$$

then its corresponding pseudoanalytic function of the second kind $\omega = B^{-1}W$ is a solution of (5.6).

Remark 83. From Proposition 82 we have also that

$$\mathcal{V} = \Pi B^{-1}.$$

Let us suppose that we have $Z_f^{(-1)}(1, z_0, z)$ and $Z_f^{(-1)}(i, z_0, z)$ where the subindex reflects the fact that these Cauchy kernels are solutions of (5.7). Then we can consider the Cauchy integral in (5.3):

$$\mathcal{C}_f W(z) = \frac{1}{2\pi} \int_\Gamma Z_f^{(-1)}(iW(\zeta)d\zeta, \zeta, z)$$

and hence we have the following Cauchy integral formula for p-analytic functions.

Theorem 84 ([72]). *Let ω be a solution of (5.5) defined in a domain Ω bounded by a simple closed continuously differentiable curve Γ and assume that ω is continuous up to the boundary Γ. Then for any $z \in \Omega$ the following equality holds:*

$$\omega(z) = B^{-1}\mathcal{C}_f B\omega(z). \tag{5.8}$$

Proof. Consider the Cauchy integral formula for solutions of (5.7), $W = \mathcal{C}_f W$ and substitute $W = B\omega$. Then application of B^{-1} gives the result.   □

5.3   The transplant operator

In this section we introduce a new concept which leads to a construction of Cauchy kernels for x^k-analytic functions. Consider the main Vekua equation (5.7) where f is a positive solution of the stationary Schrödinger equation

$$(-\Delta + \nu)f = 0 \qquad \text{in } \Omega \tag{5.9}$$

with a real-valued coefficient ν. In Theorem 33 it was shown that if W is a solution of (5.7) then its real part W_1 is necessarily a solution of (5.9), meanwhile its imaginary part W_2 is a solution of the Schrödinger equation

$$(-\Delta + \eta)h = 0 \qquad \text{in } \Omega \tag{5.10}$$

where $\eta = 2(\nabla f)^2/f^2 - \nu$. Moreover according to Theorem 36, given a solution W_1 of (5.9) the corresponding (metaharmonic conjugate) function W_2 such that $W = W_1 + iW_2$ be a solution of (5.7) can be constructed explicitly:

$$W_2 = f^{-1}\overline{A}(if^2\partial_{\overline{z}}(f^{-1}W_1)). \qquad (5.11)$$

Now let g be another positive solution of (5.9) and consider the corresponding Vekua equation

$$w_{\overline{z}} = \frac{g_{\overline{z}}}{g}\overline{w} \qquad \text{in } \Omega. \qquad (5.12)$$

We have that both $\operatorname{Re}W$ (where W is a solution of (5.7)) and $\operatorname{Re}w$ satisfy (5.9), meanwhile $\operatorname{Im}W$ and $\operatorname{Im}w$ satisfy the following, in general different Schrödinger equations,

$$(-\Delta + \eta_1)\operatorname{Im}W = 0 \qquad \text{in } \Omega$$

and

$$(-\Delta + \eta_2)\operatorname{Im}w = 0 \qquad \text{in } \Omega$$

where $\eta_1 = 2(\nabla f)^2/f^2 - \nu$ and $\eta_2 = 2(\nabla g)^2/g^2 - \nu$.

Now we introduce an operator which transforms solutions of (5.7) into solutions of (5.12) acting in the following way:

$$T_{f,g}[W] = P^+W + ig^{-1}\overline{A}(ig^2\partial_{\overline{z}}(g^{-1}P^+W)). \qquad (5.13)$$

Its application makes the imaginary part of a solution of (5.7) drop out and be substituted by an imaginary part constructed according to Theorem 36 in such a way that after this "transplant" operation the new complex function $w = T_{f,g}[W]$ becomes a solution of (5.12). This is why we call the operator $T_{f,g}$ the *transplant operator*.

Assigning a fixed value in a certain point of the domain of interest to the result of application of $\overline{A}$, we obtain an invertible one-to-one map establishing a relation between solutions of (5.7) and (5.12). The inverse to $T_{f,g}$ is given by the expression

$$T_{f,g}^{-1}[w] = T_{g,f}[w] = P^+w + if^{-1}\overline{A}(if^2\partial_{\overline{z}}(f^{-1}P^+w)).$$

Example 85. As an example of application of $T_{f,g}$ we construct the Cauchy kernel for solutions of the equation

$$w_{\overline{z}} = \frac{1}{2x}\overline{w} \qquad (5.14)$$

starting from the analytic Cauchy kernel $1/(z - z_0)$. Let $f \equiv 1$ and $g = x$. Both of them obviously satisfy the same stationary Schrödinger equation which in this case turns to be the Laplace equation. The main Vekua equation in this particular case is just the Cauchy-Riemann system $W_{\overline{z}} = 0$. The Vekua equation (5.12) takes the form (5.14). We will consider it in the right half-plane $\mathbb{C}_+$ ($\operatorname{Re}z > 0$). Now take the function $W = 1/(z - z_0)$ with $z_0 = x_0 + iy_0$ being any point in $\mathbb{C}_+$ and apply to it the transplant operator (5.13). Thus, we consider the distribution

$$w(z) = \frac{x - x_0}{|z - z_0|^2} + \frac{i}{x}\overline{A}\left(ix^2\partial_{\bar{z}}\left(\frac{x - x_0}{x\,|z - z_0|^2}\right)\right).$$

We have

$$\partial_{\bar{z}}\left(\frac{x - x_0}{x\,|z - z_0|^2}\right) = \frac{x_0}{2x^2\,|z - z_0|^2} + \frac{(x_0 - x)\,(z - z_0)}{x\,|z - z_0|^4}.$$

In order to apply the operator $\overline{A}$ to the function $ix^2\partial_{\bar{z}}\left(\frac{x-x_0}{x|z-z_0|^2}\right)$ we separate its real and imaginary parts and arrive at the equality

$$\overline{A}\left(ix^2\partial_{\bar{z}}\left(\frac{x - x_0}{x\,|z - z_0|^2}\right)\right)$$

$$= 2\left(\int_{x_1}^{x} \frac{\eta(\eta - x_0)(y - y_0)d\eta}{((\eta - x_0)^2 + (y - y_0)^2)^2} + \frac{x_0}{2}\int_{y_1}^{y}\frac{d\xi}{(x_1 - x_0)^2 + (\xi - y_0)^2}\right.$$

$$\left. -x_1(x_1 - x_0)^2\int_{y_1}^{y}\frac{d\xi}{((x_1 - x_0)^2 + (\xi - y_0)^2)^2}\right) + c_1.$$

Evaluation of the three integrals and their substitution gives us the result

$$\overline{A}\left(ix^2\partial_{\bar{z}}\left(\frac{x - x_0}{x\,|z - z_0|^2}\right)\right) = -\frac{x(y - y_0)}{|z - z_0|^2} + \arctan\frac{x - x_0}{y - y_0} + c.$$

Thus, the Cauchy kernel corresponding to the Vekua equation (5.14) has the form

$$\frac{1}{z - z_0} + \frac{i}{x}\left(\arctan\frac{x - x_0}{y - y_0} + c\right)$$

where we can choose the real constant c to be equal to zero and thus, the Cauchy kernel $Z_x^{(-1)}(1, z_0, z)$ turns out to be the distribution

$$Z_x^{(-1)}(1, z_0, z) = \frac{1}{z - z_0} + \frac{i}{x}\arctan\frac{x - x_0}{y - y_0}. \qquad (5.15)$$

5.4   Cauchy integral formulas for x^k-analytic functions

The procedure from [72] described in this section for obtaining in explicit form the Cauchy integral representations for x^k-analytic functions is based on the following simple observation.

Remark 86. For any $k \in \mathbb{R}$ the functions x^{-k} and x^{k+1} both are solutions of the Schrödinger equation

$$\left(-\Delta + \frac{k(k+1)}{x^2} \right) h(x,y) = 0.$$

Now we are ready to describe the procedure. All the equations in this section are considered in an arbitrary domain $\Omega \subset \mathbb{C}_+$ bounded by a simple closed continuously differentiable curve. Assume we have a Cauchy kernel $Z_{x^k}^{(-1)}(a, z_0; z)$ for equation (5.7) with $f = x^k$. Then $iZ_{x^k}^{(-1)}(a, z_0, z)$ is a Cauchy kernel for (5.7) with $f = x^{-k}$. Its asymptotic behavior at z_0 is $\sim ia/(z - z_0)$. That is,

$$Z_{x^{-k}}^{(-1)}(ia, z_0, z) = iZ_{x^k}^{(-1)}(a, z_0, z).$$

Consider $g = x^{k+1}$. From Remark 86 we have that between (5.7) with $f = x^{-k}$ and (5.12) with $g = x^{k+1}$ there is a relation established with the aid of the transplant operator (5.13). Thus applying $T_{f,g}$ to $iZ_{x^k}^{(-1)}(a, z_0, z)$ we obtain a Cauchy kernel for equation (5.12). Its asymptotic behavior at z_0 is again $\sim ia/(z - z_0)$. Hence

$$Z_{x^{k+1}}^{(-1)}(ia, z_0, z) = T_{x^{-k}, x^{k+1}}[iZ_{x^k}^{(-1)}(a, z_0, z)]. \tag{5.16}$$

Thus, the described procedure allows us to obtain in explicit form the Cauchy integral formula for solutions of (5.7) with $f = x^{k+1}$ from the knowledge of the Cauchy kernel corresponding to $f = x^k$.

Let us return to x^k-analytic functions. Assume k is even: $k = 2j$, $j = 0, 1, 2, \dots$. Then the coefficient in the corresponding main Vekua equation (5.7) contains $f = x^j$. As was described above the Cauchy kernel in this case can be constructed recursively from the analytic Cauchy kernel $1/(z - z_0)$. We will give more details and examples below. If k is odd: $k = 2j + 1$, $j = 0, 1, 2, \dots$, the coefficient in the corresponding main Vekua equation (5.7) contains $f = x^{j+1/2}$. The Cauchy kernel in this case $(j = 1, 2, \dots)$ can be obtained recursively from the Cauchy kernel for $f = x^{1/2}$ following the procedure described above. As was mentioned in the introduction the Cauchy kernel for $f = x^{1/2}$ is known, though it has a quite complicated form involving elliptic integrals. More precisely, this Cauchy kernel was constructed (see [2], [33]) for the Vekua equation

$$\partial_{\bar{z}} w - \frac{1}{2(z + \bar{z})}(w + \overline{w}) = 0$$

which is related to (5.7) with $f = x^{1/2}$ as follows $w = \sqrt{x}W$. We will not dwell upon the case of odd exponents k here, the purpose of this chapter is to introduce and illustrate the techniques for obtaining new Cauchy kernels from the known ones. We discuss application of our method in the case of even k, the corresponding integrals can be evaluated analytically and the Cauchy kernels can be presented in a simple and explicit form (we give examples in the case $k = 2$ and $k = 4$). Thus, we consider equation (5.7) with $f = x^j$, $j = 0, 1, 2, \ldots$. Note that the coefficient in the main Vekua equation (5.7) has the form $j/(2x)$, $j = 0, 1, 2, \ldots$. As was explained in Section 5.1, for each j in fact we need to construct two Cauchy kernels

$$Z_{x^j}^{(-1)}(1, z_0, z) \qquad \text{and} \qquad Z_{x^j}^{(-1)}(i, z_0, z).$$

In Example 85 proceeding from $1/(z - z_0)$ we constructed $Z_x^{(-1)}(1, z_0, z)$. In a similar way taking now $i/(z - z_0)$ we construct $Z_x^{(-1)}(i, z_0, z)$:

$$Z_x^{(-1)}(i, z_0, z) = T_{1,x}\left(\frac{i}{z - z_0}\right) = \frac{y - y_0}{|z - z_0|^2} + \frac{i}{x}\overline{A}\left(ix^2\partial_{\bar{z}}\left(\frac{y - y_0}{x\,|z - z_0|^2}\right)\right).$$

$$(5.17)$$

We have

$$\overline{A}\left(ix^2\partial_{\bar{z}}\left(\frac{y - y_0}{x\,|z - z_0|^2}\right)\right) = \frac{x_0(x - x_0) - (y - y_0)^2}{|z - z_0|^2} + \ln\frac{1}{|z - z_0|} + c.$$

Choosing $c = 0$ and substituting this expression into (5.17) we arrive at the result

$$Z_x^{(-1)}(i, z_0, z) = \frac{i}{z - z_0} + \frac{i}{x}\ln\frac{1}{|z - z_0|}.$$

$$(5.18)$$

Thus, from here and from (5.15) we have

$$Z_x^{(-1)}(\alpha, z_0, z) = \alpha' Z_x^{(-1)}(1, z_0, z) + \alpha'' Z_x^{(-1)}(i, z_0, z)$$

$$= \frac{\alpha}{z - z_0} + \frac{i}{x}\left(\alpha'\arctan\frac{x - x_0}{y - y_0} + \alpha''\ln\frac{1}{|z - z_0|}\right)$$

where α is an arbitrary complex number, $\alpha' = \operatorname{Re}\alpha$ and $\alpha'' = \operatorname{Im}\alpha$. We can use the freedom to add an arbitrary constant to the expression in the parentheses in order to rewrite this Cauchy kernel also in the form

$$Z_x^{(-1)}(\alpha, z_0, z) = \frac{\alpha}{z - z_0} + \frac{i}{x}\operatorname{Im}\left(\alpha\ln\frac{1}{z - z_0}\right).$$

Thus, we have obtained the Cauchy kernel for equation (5.14) which according to Theorem 84 allows us to write down the Cauchy integral formula for x^2-analytic functions. The next step is the calculation of $Z_{x^2}^{(-1)}(\alpha, z_0, z)$ using $Z_x^{(-1)}(\alpha, z_0, z)$. This will give us a Cauchy kernel for the equation

$$w_{\bar{z}} = \frac{1}{x}\overline{w}$$

$$(5.19)$$

and hence the Cauchy integral formula for x^4-analytic functions. According to (5.16) we have

$$Z_{x^2}^{(-1)}(1, z_0, z) = T_{x^{-1}, x^2}[-iZ_x^{(-1)}(i, z_0, z)] = T_{x^{-1}, x^2}\left[\frac{1}{z - z_0} + \frac{1}{x}\ln\frac{1}{|z - z_0|}\right]$$

and

$$Z_{x^2}^{(-1)}(i, z_0, z) = T_{x^{-1}, x^2}[iZ_x^{(-1)}(1, z_0, z)] = T_{x^{-1}, x^2}\left[\frac{i}{z - z_0} - \frac{1}{x}\arctan\frac{x - x_0}{y - y_0}\right].$$

We have

$$Z_{x^2}^{(-1)}(1, z_0, z) = \frac{x - x_0}{|z - z_0|^2} + \frac{1}{x}\ln\frac{1}{|z - z_0|}$$
$$+ \frac{i}{x^2}\overline{A}\left(ix^4\partial_{\bar{z}}\left(\frac{1}{x^2}\left(\frac{x - x_0}{|z - z_0|^2} + \frac{1}{x}\ln\frac{1}{|z - z_0|}\right)\right)\right) \qquad (5.20)$$

and

$$Z_{x^2}^{(-1)}(i, z_0, z) = \frac{y - y_0}{|z - z_0|^2} - \frac{1}{x}\arctan\frac{x - x_0}{y - y_0}$$
$$+ \frac{i}{x^2}\overline{A}\left(ix^4\partial_{\bar{z}}\left(\frac{1}{x^2}\left(\frac{y - y_0}{|z - z_0|^2} - \frac{1}{x}\arctan\frac{x - x_0}{y - y_0}\right)\right)\right). \qquad (5.21)$$

Evaluation of the corresponding integrals gives us the result

$$\overline{A}\left(ix^4\partial_{\bar{z}}\left(\frac{1}{x^2}\left(\frac{x - x_0}{|z - z_0|^2} + \frac{1}{x}\ln\frac{1}{|z - z_0|}\right)\right)\right)$$
$$= 3(y - y_0)(\ln|z - z_0| - 1) + 3x_0\arctan\frac{x - x_0}{y - y_0} - \frac{x^2(y - y_0)}{|z - z_0|^2} + c_1$$

and

$$\overline{A}\left(ix^4\partial_{\bar{z}}\left(\frac{1}{x^2}\left(\frac{y - y_0}{|z - z_0|^2} - \frac{1}{x}\arctan\frac{x - x_0}{y - y_0}\right)\right)\right)$$
$$= -3x_0\ln|z - z_0| - 3x + \frac{x^2(x - x_0)}{|z - z_0|^2} + 3(y - y_0)\arctan\frac{x - x_0}{y - y_0} + c_2.$$

Choosing $c_1 = c_2 = 0$ and substituting these expressions into (5.20) and (5.21) we obtain

$$Z_{x^2}^{(-1)}(1, z_0, z) = \frac{1}{z - z_0} + \frac{1}{x}\left(1 - \frac{3i(y - y_0)}{x}\right)\ln\frac{1}{|z - z_0|}$$
$$- \frac{3iy}{x^2} + \frac{3ix_0}{x^2}\arctan\frac{x - x_0}{y - y_0}$$

and

$$Z_{x^2}^{(-1)}(i, z_0, z) = \frac{i}{z - z_0} - \frac{1}{x}\left(1 - \frac{3i(y - y_0)}{x}\right)\arctan\frac{x - x_0}{y - y_0}$$
$$- \frac{3i}{x} + \frac{3ix_0}{x^2}\ln\frac{1}{|z - z_0|}.$$

It is not difficult to verify by a direct substitution that both functions are indeed solutions of (5.19) in $\mathbb{C}_+ \setminus z_0$. Combining them we can write down the Cauchy kernel for the Vekua equation (5.19) in the form

$$Z_{x^2}^{(-1)}(\alpha, z_0, z) = \alpha' Z_{x^2}^{(-1)}(1, z_0, z) + \alpha'' Z_{x^2}^{(-1)}(i, z_0, z)$$
$$= \frac{\alpha}{z - z_0} + \frac{3ix_0}{x^2}\operatorname{Im}\left(\alpha\ln\frac{1}{z - z_0}\right) - \frac{3i}{x^2}\operatorname{Im}(\alpha z)$$
$$+ \frac{1}{x}\left(1 - \frac{3i(y - y_0)}{x}\right)\operatorname{Re}\left(\alpha\ln\frac{1}{z - z_0}\right).$$

The constructed Cauchy kernel gives us the Cauchy integral formula for x^4-analytic functions (see Theorem 84). The procedure can be continued and the Cauchy integral representations for x^j-analytic functions can be obtained for higher exponents j. It is sufficiently simple and can be implemented with the aid of a package of symbolic calculation.

Chapter 6

Complex Riccati Equation

6.1 Preliminary notes

The ordinary differential Riccati equation

$$u' = pu^2 + qu + r \tag{6.1}$$

has received a great deal of attention since a particular version was first studied by Count Riccati in 1724, owing to both its specific properties and the wide range of applications in which it appears. For a survey of the history and classical results on this equation, see for example [34] and [107]. This equation can be always reduced to its canonical form (see, e.g., [18], [56]),

$$y' + y^2 = v, \tag{6.2}$$

and this is the form that we will consider.

One of the reasons for which the Riccati equation has so many applications is that it is related to the general second-order homogeneous differential equation. In particular, the one-dimensional Schrödinger equation

$$-\frac{d^2}{dx^2}u + v(x)u = 0 \tag{6.3}$$

is related to (6.2) by the easily inverted substitution

$$y = \frac{u'}{u}, \qquad u = e^{\int y}.$$

This substitution, which as its most spectacular application reduces Burger's equation to the standard one-dimensional heat equation, is the basis of the well-developed theory of logarithmic derivatives for the integration of nonlinear differential equations [89]. A generalization of this substitution will be used in this work.

A second relation between the one-dimensional Schrödinger equation and the Riccati equation is as follows. The one-dimensional Schrödinger operator can be factorized in the form

$$-\frac{d^2}{dx^2} + v(x) = -\left(\frac{d}{dx} + y(x)\right)\left(\frac{d}{dx} - y(x)\right) \tag{6.4}$$

if and only if (6.2) holds. This observation is a key to a vast area of research related to the factorization method (see, e.g., [93] and [94]) and to the Darboux transformation (see, e.g., [90], [97] and [112]). Here we consider a result similar to (6.4) but already in a two-dimensional setting.

Among the peculiar properties of the Riccati equation stand out two theorems of Euler, dating from 1760. The first of these states that if a particular solution y_0 of the Riccati equation is known, the substitution $y = y_0 + z$ reduces (6.2) to a Bernoulli equation which in turn is reduced by the substitution $z = \frac{1}{u}$ to a first-order linear equation. Thus given a particular solution of the Riccati equation, it can be linearized and the general solution can be found in two integrations. The second of these theorems states that given two particular solutions y_0, y_1 of the Riccati equation, the general solution can be found in the form

$$y = \frac{ky_0 \exp(\int y_0 - y_1) - y_1}{k \exp(\int y_0 - y_1) - 1} \tag{6.5}$$

where k is a constant. That is, given two particular solutions of (6.2), the general solution can be found in one integration.

Other interesting properties are those discovered by Weyr and Picard ([122], [34]). The first is that given a third particular solution y_3, the general solution can be found without integrating. That is, an explicit combination of three particular solutions gives the general solution. The second is that given a fourth particular solution y_4, the cross ratio

$$\frac{(y_1 - y_2)(y_3 - y_4)}{(y_1 - y_4)(y_3 - y_2)}$$

is a constant. Thus the derivative of this ratio is zero, which holds if and only if the numerator of the derivative is zero, that is, if and only if

$$(y_1 - y_4)(y_3 - y_2)((y_1 - y_2)(y_3 - y_4))' - (y_1 - y_2)(y_3 - y_4)((y_1 - y_4)(y_3 - y_2))' = 0.$$

Dividing by $(y_1 - y_2)(y_3 - y_4)(y_1 - y_4)(y_3 - y_2)$, we see that Picard's theorem is equivalent to the statement

$$\frac{(y_1 - y_2)'}{y_1 - y_2} + \frac{(y_3 - y_4)'}{y_3 - y_4} - \frac{(y_1 - y_4)'}{y_1 - y_4} - \frac{(y_3 - y_2)'}{y_3 - y_2} = 0. \tag{6.6}$$

In this chapter we study the equation

$$\partial_{\bar{z}}Q + |Q|^2 = v \tag{6.7}$$

where z is a complex variable, $\partial_{\bar{z}} = \frac{1}{2}(\partial_x + i\partial_y)$, Q is a complex-valued function of z and v is a real-valued function. Note that this equation is different from the complex Riccati equation studied in dozens of works where it is supposed to have the form (6.1) with complex analytic coefficients p, q and r and a complex analytic solution u (see, e.g., [53]). We do not suppose analyticity of the complex functions involved and show that equation (6.7), unlike the equation considered in [53], is related to the two-dimensional stationary Schrödinger equation in a similar way as the ordinary Riccati and Schrödinger equations are related in dimension 1. Moreover, we establish generalizations of the Euler and Picard theorems and obtain some other results which are essentially two-dimensional, for example, an analogue of the Cauchy integral theorem for solutions of the complex Riccati equation (6.7).

Equation (6.7) first appeared in [68] as a reduction to a two-dimensional case of the spatial factorization of the stationary Schrödinger operator which was studied in a quaternionic setting in [9], [11], [63], [64], [66], [68] and later on using Clifford analysis in [10] and [36].

The ordinary Riccati equation is at the heart of many analytical and numerical approaches to problems involving the one-dimensional Schrödinger and Sturm-Liouville equations. Here we furnish a complete structural analogy between dimensions 1 and 2 regarding the relationship between the Schrödinger and the Riccati equations. Besides, the deep similarity between the ordinary Riccati equation and (6.7) strongly suggests that many known applications of the ordinary Riccati equation can be generalized to the two-dimensional situation and many new aspects such as Theorem 94 will become manifest.

The results discussed in this chapter have been obtained together with K.V. Khmelnytskaya and presented in [60].

6.2 The two-dimensional stationary Schrödinger equation and the complex Riccati equation

Consider the complex differential Riccati equation

$$\partial_{\bar{z}}Q + |Q|^2 = \frac{\nu}{4} \tag{6.8}$$

where for convenience the factor $1/4$ was included. We recall that ν is a real-valued function. Together with this equation we consider the stationary Schrödinger equation

$$(-\Delta + \nu)u = 0 \tag{6.9}$$

where u is real-valued. Both equations are studied in a domain $\Omega \subset \mathbb{C}$.

Theorem 87. *Let u be a solution of* (6.9). *Then its logarithmic derivative*

$$Q = \frac{u_z}{u} \tag{6.10}$$

is a solution of (6.8).

Proof. It is only necessary to substitute (6.10) into (6.8).   □

Remark 88. Any solution of equation (6.8) fulfils (2.17). Indeed, the imaginary part of (6.8) reads

$$\partial_y Q_1 + \partial_x Q_2 = 0.$$

Theorem 89. *Let Q be a solution of* (6.8)*. Then the function*

$$u = e^{A[Q]} \tag{6.11}$$

is a solution of (6.9)*.*

Proof. Equation (6.9) can be written in the form

$$(4\partial_{\bar{z}}\partial_z - \nu)u = 0.$$

Taking u in the form (6.11) where Q is a solution of (6.8) and using the observation that

$$\partial_{\bar{z}}(A[Q]) = \overline{\partial_z(A[Q])} = \overline{Q},$$

we have

$$\partial_{\bar{z}}\partial_z u = \partial_{\bar{z}}\left(Qe^{A[Q]}\right) = e^{A[Q]}\left(\partial_{\bar{z}}Q + |Q|^2\right) = \frac{\nu}{4}u. \qquad \square$$

Observe that this theorem means that if Q is a solution of (6.8) then there exists a solution u of (6.9) such that (6.10) is valid. Theorems 87 and 89 are direct generalizations of the corresponding facts from the one-dimensional theory.

The following statement is a generalization of the one-dimensional factorization (6.4).

Theorem 90. *Given a complex function Q, for any real-valued twice-continuously differentiable function φ the following equality is valid:*

$$\frac{1}{4}\left(\Delta - \nu\right)\varphi = (\partial_{\bar{z}} + QC)(\partial_z - QC)\varphi \tag{6.12}$$

$$= (\partial_z + \overline{Q}C)(\partial_{\bar{z}} - \overline{Q}C)\varphi$$

if and only if Q is a solution of the Riccati equation (6.8)*.*

Proof. Consider

$$(\partial_{\bar{z}} + QC)(\partial_z - QC)\varphi = \frac{1}{4}\Delta\varphi - |Q|^2\,\varphi - Q_{\bar{z}}\varphi$$

from which it is seen that (6.12) is valid iff Q is a solution of (6.8). The second equality in (6.12) is obtained by applying C to both sides of the first equality.   □

6.3   Generalizations of classical theorems

In this section we give generalizations of both Euler's theorems for the Riccati equation as well as of Picard's theorem.

Theorem 91 (First Euler's theorem). *Let Q_0 be a bounded particular solution of (6.8). Then (6.8) reduces to the following first-order (real-linear) equation*

$$W_{\bar{z}} = \overline{Q_0 W} \tag{6.13}$$

in the following sense. Any solution of (6.8) has the form

$$Q = \frac{\partial_z \operatorname{Re} W}{\operatorname{Re} W}$$

and vice versa, any solution of (6.13) can be expressed via a corresponding solution Q of (6.8) as

$$W = e^{A[Q]} + ie^{-A[Q_0]}\overline{A}\left[ie^{2A[Q_0]}\partial_{\bar{z}}e^{A[Q-Q_0]}\right]. \tag{6.14}$$

Proof. Let Q_0 be a bounded solution of (6.8). Then by Theorem 89 we have that there exists a nonvanishing real-valued solution f of (6.9) such that $Q_0 = f_z/f$. Hence (6.13) has the form (3.15). Now, let Q be any solution of (6.8). Then again $Q = u_z/u$ where u is a solution of (6.9). According to Theorem 36, u is a real part of a solution W of (3.15). Thus we have proved the first part of the theorem.

Let $W = u + iv$ be any solution of (6.13) ($u = \operatorname{Re} W$). Then u is a solution of (6.9) and by Theorem 87 it can be represented in the form $u = e^{A[Q]}$ where Q is a solution of (6.8). Then by Theorem 36 (formula (3.26)), W has the form (6.14). $\qquad\square$

Thus the Riccati equation (6.8) is equivalent to a main Vekua equation of the form (3.15).

In what follows we suppose that in the domain of interest Ω there exists a bounded solution of (6.8).

Theorem 92 (Second Euler's theorem). *Any solution Q of equation (6.8) defined for $|z - z_0| < R$ can be represented in the form*

$$Q = \frac{\sum_{n=0}^{\infty} Q_n e^{A[Q_n]}}{\sum_{n=0}^{\infty} e^{A[Q_n]}} \tag{6.15}$$

where $\{Q_n\}_{n=0}^{\infty}$ is the set of particular solutions of the Riccati equation (6.8) obtained as follows:

$$Q_n(z) = \frac{\partial_z \operatorname{Re} Z^{(n)}(a_n, z_0, z)}{\operatorname{Re} Z^{(n)}(a_n, z_0, z)},$$

$Z^{(n)}(a_n, z_0, z)$ are formal powers corresponding to equation (6.13) and both series in (6.15) converge normally for $|z - z_0| < R$.

Proof. By the first Euler theorem we have

$$Q = \frac{\partial_z \operatorname{Re} W}{\operatorname{Re} W}$$

where W is a solution of (6.13). From Theorem 65 we obtain

$$Q(z) = \frac{\partial_z \sum_{n=0}^{\infty} \operatorname{Re} Z^{(n)}(a_n, z_0; z)}{\sum_{n=0}^{\infty} \operatorname{Re} Z^{(n)}(a_n, z_0; z)}.$$

Every formal power $Z^{(n)}(a_n, z_0; z)$ corresponds to a solution of (6.8):

$$Q_n(z) = \frac{\partial_z \operatorname{Re} Z^{(n)}(a_n, z_0; z)}{\operatorname{Re} Z^{(n)}(a_n, z_0; z)}$$

or $\operatorname{Re} Z^{(n)}(a_n, z_0; z) = e^{A[Q_n](z)}$ from where we obtain (6.15).   □

In the next statement we give a generalization of Picard's theorem in the form (6.6).

Theorem 93 (Picard's theorem). *Let Q_k, $k = 1, 2, 3, 4$ be four solutions of* (6.8). *Then*

$$\frac{\partial_{\bar{z}}(Q_1 - Q_2) + 2i \operatorname{Im}(\overline{Q}_1 Q_2)}{Q_1 - Q_2} + \frac{\partial_{\bar{z}}(Q_3 - Q_4) + 2i \operatorname{Im}(\overline{Q}_3 Q_4)}{Q_3 - Q_4}$$
$$- \frac{\partial_{\bar{z}}(Q_1 - Q_4) + 2i \operatorname{Im}(\overline{Q}_1 Q_4)}{Q_1 - Q_4} - \frac{\partial_{\bar{z}}(Q_3 - Q_2) + 2i \operatorname{Im}(\overline{Q}_3 Q_2)}{Q_3 - Q_2} = 0.$$

Proof. Obviously,

$$(\overline{Q}_1 + \overline{Q}_4) + (\overline{Q}_3 + \overline{Q}_2) - (\overline{Q}_1 + \overline{Q}_2) - (\overline{Q}_3 + \overline{Q}_4) = 0.$$

Multiplying each parenthesis by $1 = (Q_i - Q_j)/(Q_i - Q_j)$ we obtain the equality

$$\frac{(\overline{Q}_1 + \overline{Q}_4)(Q_1 - Q_4)}{(Q_1 - Q_4)} + \frac{(\overline{Q}_3 + \overline{Q}_2)(Q_3 - Q_2)}{(Q_3 - Q_2)}$$
$$- \frac{(\overline{Q}_1 + \overline{Q}_2)(Q_1 - Q_2)}{(Q_1 - Q_2)} - \frac{(\overline{Q}_3 + \overline{Q}_4)(Q_3 - Q_4)}{(Q_3 - Q_4)} = 0.$$

Using

$$(\overline{Q}_i + \overline{Q}_j)(Q_i - Q_j) = \partial_{\bar{z}}(Q_j - Q_i) - \overline{Q}_i Q_j + Q_i \overline{Q}_j$$

the result is obtained.   □

6.4 Cauchy's integral theorem

Theorem 94 (Cauchy's integral theorem for the complex Riccati equation). *Let Q_0 and Q_1 be bounded solutions of (6.8) in Ω. Then for every closed curve Γ lying in a simply connected subdomain of Ω,*

$$\operatorname{Re} \int_\Gamma (Q_1 - Q_0)\, e^{A[Q_1 - Q_0]} dz + i \operatorname{Im} \int_\Gamma (Q_1 - Q_0)\, e^{A[Q_1 + Q_0]} dz = 0. \tag{6.16}$$

Proof. From Theorem 89 we have that $f = e^{A[Q_0]}$ and $u = e^{A[Q_1]}$ are solutions of (6.9). Now, applying Theorem 43 we obtain

$$\operatorname{Re} \int_\Gamma \partial_z(\frac{u}{f})dz + i \operatorname{Im} \int_\Gamma f^2 \partial_z(\frac{u}{f})dz = 0$$

for every closed curve Γ situated in a simply connected subdomain of Ω, which gives us the equality

$$\operatorname{Re} \int_\Gamma \partial_z(e^{A[Q_1 - Q_0]})dz + i \operatorname{Im} \int_\Gamma e^{2A[Q_0]} \partial_z(e^{A[Q_1 - Q_0]})dz = 0.$$

From this we obtain the result. □

As a particular case let us analyze the Riccati equation (6.8) with $\nu \equiv 0$ which is related to the Laplace equation. If in (6.16) we assume that $Q_0 \equiv 0$, then (6.16) takes the form

$$\int_\Gamma Q_1 e^{A[Q_1]} dz = 0.$$

This is obviously valid because if Q_1 is another bounded solution of the Riccati equation with $\nu \equiv 0$, then according to Theorem 89 we have that $u = e^{A[Q_1]}$ is a harmonic function and the last formula turns into the equality

$$\int_\Gamma u_z dz = 0 \tag{6.17}$$

(u_z is analytic).

Now, if in (6.16) we assume that $Q_1 \equiv 0$, then (6.16) takes the form

$$\operatorname{Re} \int_\Gamma Q_0 e^{-A[Q_0]} dz + i \operatorname{Im} \int_\Gamma Q_0 e^{A[Q_0]} dz = 0.$$

Rewriting this equality in terms of the harmonic function $f = e^{A[Q_0]}$ we obtain

$$\operatorname{Re} \int_\Gamma \frac{f_z dz}{f^2} + i \operatorname{Im} \int_\Gamma f_z dz = 0$$

which taking into account (6.17) becomes the equality

$$\operatorname{Re} \int_\Gamma \frac{f_z dz}{f^2} = 0$$

or which is the same,

$$\text{Re} \int_\Gamma \partial_z \left(\frac{1}{f} \right) dz = 0. \tag{6.18}$$

This equality is a simple corollary of a complex version of the Green-Gauss theorem (see, e.g., [117, sect. 3.2]) according to which we have

$$\frac{1}{2i} \int_\Gamma \partial_z \left(\frac{1}{f} \right) dz = \int_\Omega \partial_{\bar{z}} \partial_z \left(\frac{1}{f} \right) dxdy.$$

For f real the right-hand side is real-valued and we obtain (6.18).

Part II

Applications to Sturm-Liouville Theory

Chapter 7

A Representation for Solutions of the Sturm-Liouville Equation

The Sturm-Liouville equation

$$(pu')' + qu = \omega^2 u$$

is one of the canonical and main objects of study in classical mathematical physics. Its theory has a long history and is quite well developed (see, e.g., [86], [129]).

This part of the book is dedicated to a new representation for solutions of the Sturm-Liouville equation which was obtained in [74] with the aid of pseudoanalytic function theory and allows us to reduce corresponding spectral problems to a problem of finding zeros of analytic functions. The fact that a Sturm-Liouville problem is equivalent to finding zeros of a corresponding analytic function has been known since long ago. Nevertheless, given a spectral Sturm-Liouville problem, how to construct the associated analytic function of ω whose zeros coincide with the eigenvalues was an open question. Here we show how the Taylor coefficients of this analytic function can be obtained explicitly and discuss different applications including a numerical method based on this result.

The main result of this part is formulated in Section 7.8, before we explain how it is obtained using the results of the preceding chapters. In Chapter 8 we discuss some applications of the main result including spectral problems and the Darboux transformation.

7.1 Solving the one-dimensional Schrödinger equation

Consider the equation

$$(-\partial_x^2 + q(x))g(x) = 0. \tag{7.1}$$

We suppose that q and g are real-valued functions and that on some interval I_x of the independent variable x there exists a bounded nonvanishing solution

$g_0 \in C^2(I_x)$ such that $1/g_0$ is also bounded. For simplicity we suppose that the interval I_x includes the point $x = 0$ and that $g_0(0) = 1$. Our first goal is to solve the equation

$$(-\partial_x^2 + q(x) \pm \omega^2)u(x) = 0$$

for any real constant ω. We start with the "+"-case.

7.2   The "+"-case

Consider the equation

$$(-\partial_x^2 + q(x) + \omega^2)u(x) = 0. \tag{7.2}$$

Let us notice that for the equation

$$(-\Delta + q(x) + \omega^2)U(x, y) = 0 \tag{7.3}$$

where $\Delta = \partial_x^2 + \partial_y^2$ we can immediately propose a particular solution, e.g.,

$$f(x, y) = g_0(x)e^{\omega y}. \tag{7.4}$$

This function does not have zeros on any rectangular domain $\Omega = I_x \times I_y$ where I_y is an arbitrary finite interval of the variable y. For simplicity we assume that the origin $z = 0$ is an internal point of the domain Ω. According to Theorem 36, any solution of (7.3) is a real part of a solution of the main Vekua equation (3.15) where f is defined by (7.4). Moreover, a generating pair for this Vekua equation has the form

$$F(x, y) = f(x, y) = g_0(x)e^{\omega y}$$

and

$$G(x, y) = i/f(x, y) = ig_0^{-1}(x)e^{-\omega y}.$$

Now using the results from Subsection 4.2 we can construct the formal powers corresponding to equation (3.15) and to this generating pair. We have that

$$Z^{(n)}(a, 0, z) = g_0(x)e^{\omega y} \operatorname{Re} {}_*Z^{(n)}(a, 0, z) + ig_0^{-1}(x)e^{-\omega y} \operatorname{Im} {}_*Z^{(n)}(a, 0, z)$$

where ${}_*Z^{(n)}(a, 0, z)$ are constructed according to formulas (4.5) and (4.6) where $\sigma(x) = g_0(x)$ and $\tau(y) = e^{-\omega y}$.

We know that any solution W of (3.15) in Ω can be represented in the form

$$W(z) = \sum_{n=0}^{\infty} Z^{(n)}(a_n, 0, z)$$

and hence any solution U of (7.3) has the form

$$U(x, y) = \sum_{n=0}^{\infty} \operatorname{Re} Z^{(n)}(a_n, 0, z) \tag{7.5}$$

$$= g_0(x)e^{\omega y} \sum_{n=0}^{\infty} \left(a_n' \operatorname{Re} {}_*Z^{(n)}(1, 0, z) + a_n'' \operatorname{Re} {}_*Z^{(n)}(i, 0, z) \right).$$

Observe that solutions of (7.2) are also solutions of (7.3). Consequently, for any solution u of (7.2) there exists a set of real numbers $\{a'_n, a''_n\}_{n=0}^{\infty}$ such that

$$u(x) = g_0(x)e^{\omega y} \sum_{n=0}^{\infty} \left(a'_n \operatorname{Re} {}_*Z^{(n)}(1,0,z) + a''_n \operatorname{Re} {}_*Z^{(n)}(i,0,z) \right). \tag{7.6}$$

In other words there exist such sets of coefficients $\{a'_n, a''_n\}_{n=0}^{\infty}$ that

$$\partial_y \left(e^{\omega y} \sum_{n=0}^{\infty} \left(a'_n \operatorname{Re} {}_*Z^{(n)}(1,0,z) + a''_n \operatorname{Re} {}_*Z^{(n)}(i,0,z) \right) \right) \equiv 0. \tag{7.7}$$

Obviously, if this condition is fulfilled, the resulting function (7.6) is a solution of (7.2).

Let us analyze equation (7.7). First of all we have that, for an odd n,

$$_*Z^{(n)}(1,0,z) = \sum_{j=0}^{n} \binom{n}{j} X^{(n-j)} i^j Y^{(j)},$$

$$_*Z^{(n)}(i,0,z) = i\sum_{j=0}^{n} \binom{n}{j} \widetilde{X}^{(n-j)} i^j \widetilde{Y}^{(j)}$$

and for an even n,

$$_*Z^{(n)}(1,0,z) = \sum_{j=0}^{n} \binom{n}{j} \widetilde{X}^{(n-j)} i^j Y^{(j)},$$

$$_*Z^{(n)}(i,0,z) = i\sum_{j=0}^{n} \binom{n}{j} X^{(n-j)} i^j \widetilde{Y}^{(j)}.$$

Thus, we obtain for an odd n,

$$\operatorname{Re} {}_*Z^{(n)}(1,0,z) = \sum_{\text{even } j=0}^{n} \binom{n}{j} X^{(n-j)} i^j Y^{(j)},$$

$$\operatorname{Re} {}_*Z^{(n)}(i,0,z) = \sum_{\text{odd } j=1}^{n} \binom{n}{j} \widetilde{X}^{(n-j)} i^{j+1} \widetilde{Y}^{(j)}$$

and for an even n,

$$\operatorname{Re} {}_*Z^{(n)}(1,0,z) = \sum_{\text{even } j=0}^{n} \binom{n}{j} \widetilde{X}^{(n-j)} i^j Y^{(j)},$$

$$\operatorname{Re} {}_*Z^{(n)}(i,0,z) = \sum_{\text{odd } j=1}^{n} \binom{n}{j} X^{(n-j)} i^{j+1} \widetilde{Y}^{(j)}.$$

It is somewhat more convenient for what follows to rewrite these formulas in the following equivalent form

$$
\begin{cases}
\operatorname{Re} {}_*Z^{(n)}(1,0,z) = \displaystyle\sum_{\text{odd } k=1}^{n} \binom{n}{k} X^{(k)} i^{n-k} Y^{(n-k)}, \\[4mm]
\operatorname{Re} {}_*Z^{(n)}(i,0,z) = \displaystyle\sum_{\text{even } k=0}^{n} \binom{n}{k} \widetilde{X}^{(k)} i^{n-k+1} \widetilde{Y}^{(n-k)}
\end{cases}
\qquad \text{for an odd } n \quad (7.8)
$$

and

$$
\begin{cases}
\operatorname{Re} {}_*Z^{(n)}(1,0,z) = \displaystyle\sum_{\text{even } k=0}^{n} \binom{n}{k} \widetilde{X}^{(k)} i^{n-k} Y^{(n-k)}, \\[4mm]
\operatorname{Re} {}_*Z^{(n)}(i,0,z) = \displaystyle\sum_{\text{odd } k=1}^{n} \binom{n}{k} X^{(k)} i^{n-k+1} \widetilde{Y}^{(n-k)}
\end{cases}
\qquad \text{for an even } n. \quad (7.9)
$$

We recall that

$$
Y^{(n)}(y) =
\begin{cases}
n \displaystyle\int_0^y Y^{(n-1)}(\eta) e^{2\omega\eta} d\eta & \text{for an odd } n, \\[4mm]
n \displaystyle\int_0^y Y^{(n-1)}(\eta) e^{-2\omega\eta} d\eta & \text{for an even } n,
\end{cases}
$$

$$
\widetilde{Y}^{(n)}(y) =
\begin{cases}
n \displaystyle\int_0^y \widetilde{Y}^{(n-1)}(\eta) e^{-2\omega\eta} d\eta & \text{for an odd } n, \\[4mm]
n \displaystyle\int_0^y \widetilde{Y}^{(n-1)}(\eta) e^{2\omega\eta} d\eta & \text{for an even } n.
\end{cases}
$$

Thus, from (7.8) and (7.9) we have

$$
\begin{cases}
\partial_y \operatorname{Re} {}_*Z^{(n)}(1,0,z) = e^{-2\omega y} \displaystyle\sum_{\text{odd } k=1}^{n} (n-k)\binom{n}{k} X^{(k)} i^{n-k} Y^{(n-k-1)}, \\[4mm]
\partial_y \operatorname{Re} {}_*Z^{(n)}(i,0,z) = e^{-2\omega y} \displaystyle\sum_{\text{even } k=0}^{n} (n-k)\binom{n}{k} \widetilde{X}^{(k)} i^{n-k+1} \widetilde{Y}^{(n-k-1)}
\end{cases}
$$

$$
\text{for an odd } n \quad (7.10)
$$

and

$$
\begin{cases}
\partial_y \operatorname{Re} {}_*Z^{(n)}(1,0,z) = e^{-2\omega y} \displaystyle\sum_{\text{even } k=0}^{n} (n-k)\binom{n}{k} \widetilde{X}^{(k)} i^{n-k} Y^{(n-k-1)}, \\[4mm]
\partial_y \operatorname{Re} {}_*Z^{(n)}(i,0,z) = e^{-2\omega y} \displaystyle\sum_{\text{odd } k=1}^{n} (n-k)\binom{n}{k} X^{(k)} i^{n-k+1} \widetilde{Y}^{(n-k-1)}
\end{cases}
$$

$$
\text{for an even } n > 0. \quad (7.11)
$$

Now returning to equation (7.7) we observe that it is equivalent to the equation

$$\sum_{n=0}^{\infty} \Big(a_n' \big(\omega \operatorname{Re} {}_* Z^{(n)}(1,0,z) + \partial_y \operatorname{Re} {}_* Z^{(n)}(1,0,z) \big)$$
$$+ a_n'' \big(\omega \operatorname{Re} {}_* Z^{(n)}(i,0,z) + \partial_y \operatorname{Re} {}_* Z^{(n)}(i,0,z) \big) \Big) = 0.$$

from which using (7.8), (7.9) and (7.10), (7.11) we obtain that (7.7) can be written as

$$a_0' \omega + \sum_{\substack{n=2 \\ \text{even}}}^{\infty} \Big(a_n' \sum_{\substack{k=0 \\ \text{even}}}^{n} i^{n-k} \binom{n}{k} \widetilde{X}^{(k)} (\omega Y^{(n-k)} + (n-k) Y^{(n-k-1)} e^{-2\omega y})$$

$$+ a_n'' \sum_{\substack{k=1 \\ \text{odd}}}^{n} i^{n-k+1} \binom{n}{k} X^{(k)} (\omega \widetilde{Y}^{(n-k)} + (n-k) \widetilde{Y}^{(n-k-1)} e^{-2\omega y}))$$

$$+ \sum_{\substack{n=1 \\ \text{odd}}}^{\infty} \Big(a_n' \sum_{\substack{k=1 \\ \text{odd}}}^{n} i^{n-k} \binom{n}{k} X^{(k)} (\omega Y^{(n-k)} + (n-k) Y^{(n-k-1)} e^{-2\omega y})$$

$$+ a_n'' \sum_{\substack{k=0 \\ \text{even}}}^{n} i^{n-k+1} \binom{n}{k} \widetilde{X}^{(k)} (\omega \widetilde{Y}^{(n-k)} + (n-k) \widetilde{Y}^{(n-k-1)} e^{-2\omega y})) = 0.$$

$$(7.12)$$

In order that this equality hold identically, the expressions corresponding to different $X^{(n)}$ and $\widetilde{X}^{(n)}$ for all n should vanish identically. Combining all terms multiplied by $\widetilde{X}^{(0)}$ we obtain the equation

$$a_0' \omega + \sum_{\substack{n=2 \\ \text{even}}}^{\infty} a_n' i^n (\omega Y^{(n)} + n Y^{(n-1)} e^{-2\omega y})$$

$$+ \sum_{\substack{n=1 \\ \text{odd}}}^{\infty} a_n'' i^{n+1} (\omega \widetilde{Y}^{(n)} + n \widetilde{Y}^{(n-1)} e^{-2\omega y}) = 0. \qquad (7.13)$$

Gathering all terms multiplied by $X^{(1)}$ we obtain the second equation

$$\sum_{\substack{n=2 \\ \text{even}}}^{\infty} a_n'' i^n n (\omega \widetilde{Y}^{(n-1)} + (n-1) \widetilde{Y}^{(n-2)} e^{-2\omega y})$$

$$+ \sum_{\substack{n=1 \\ \text{odd}}}^{\infty} a_n' i^{n-1} n (\omega Y^{(n-1)} + (n-1) Y^{(n-2)} e^{-2\omega y}) = 0$$

which can be rewritten as

$$a_1' \omega + \sum_{\substack{n=2 \\ \text{even}}}^{\infty} a_{n+1}' i^n (n+1) (\omega Y^{(n)} + n Y^{(n-1)} e^{-2\omega y})$$

$$+ \sum_{\substack{n=1 \\ \text{odd}}}^{\infty} a_{n+1}'' i^{n+1} (n+1) (\omega \widetilde{Y}^{(n)} + n \widetilde{Y}^{(n-1)} e^{-2\omega y}) = 0. \qquad (7.14)$$

Gathering all terms multiplied by $\widetilde{X}^{(2)}$, $X^{(3)}$, ... we obtain an infinite system of equations which fortunately we do not need to solve. Here we are reasoning along the following lines. First of all we observe that if such sets of coefficients $\{a'_n, a''_n\}_{n=0}^{\infty}$ exist that all the equations derived from (7.12) are satisfied, they do not depend on functions $X^{(n)}$ and $\widetilde{X}^{(n)}$ but only on $Y^{(n)}$ and $\widetilde{Y}^{(n)}$. Thus, they can be constructed independently of the concrete form of g_0 and hence of the potential q. Second, we know that such sets of coefficients exist. This is due to our earlier observation that solutions of (7.2) are also solutions of (7.3) and hence they can be written in the form (7.5).

These two arguments lead to the following surprising solution. We can take any q, for example, $q \equiv 0$ and any pair of independent solutions of the resulting Schrödinger equation (7.2) and to obtain their corresponding sets of coefficients. These two sets will be universal in the sense that the general solution of (7.2) for any other q will be constructed with the aid of this pair of sets of coefficients just by changing the generating function g_0 and obtaining a corresponding system of functions $X^{(n)}$ and $\widetilde{X}^{(n)}$. On first glance this conclusion can appear against intuition, nevertheless its more detailed analysis as well as the final result convince us that it is really natural. Thus, in the next subsection we construct such a pair of sets of Taylor coefficients (in formal powers).

7.3 Two sets of Taylor coefficients

Here we consider the case $q \equiv 0$. Then the Schrödinger equation (7.2) becomes

$$(-\partial_x^2 + \omega^2)u(x) = 0. \tag{7.15}$$

Note that the corresponding equation (7.1) has the form $\partial_x^2 g(x) = 0$ and possesses a suitable particular solution satisfying all the requirements (see the beginning of Section 7) $g_0 \equiv 1$. Then $f = e^{\omega y}$ and the main Vekua equation in this case has the form

$$W_{\bar{z}} = \frac{i\omega}{2}\overline{W}. \tag{7.16}$$

Let us take two independent solutions of (7.15) $u^+(x) = e^{\omega x}$ and $u^-(x) = e^{-\omega x}$. First, we obtain the set of coefficients $\{a'_n, a''_n\}_{n=0}^{\infty}$ for the function u^+. The first step consists in constructing the corresponding conjugate metaharmonic function v^+ (see Theorem 36)

$$v^+ = e^{-\omega y}\overline{A}(ie^{2\omega y}\partial_{\bar{z}}e^{\omega(x-y)}) = e^{-\omega y}\overline{A}(\frac{\omega}{2}(1+i)e^{\omega(x+y)}).$$

We have

$$\overline{A}(\frac{\omega}{2}(1+i)e^{\omega(x+y)}) = e^{\omega(x+y)} + c.$$

We choose $c = 0$, then $v^+ = e^{\omega x}$. Thus, one of the solutions of the main Vekua equation (7.16) such that $u^+ = e^{\omega x}$ is its real part has the form

$$W^+ = (1+i)e^{\omega x}. \tag{7.17}$$

Now, in order to construct its corresponding Taylor coefficients (in formal powers) we notice that $A_{(F,G)} = 0$ and $B_{(F,G)} = -i\omega/2$ (here $F = e^{\omega y}$ and $G = ie^{-\omega y}$). Thus, the operation of the (F, G)-derivative has the form

$$\dot{W} = W_z + \frac{i\omega}{2}\overline{W}.$$

For the function (7.17) we have

$$\dot{W}^+ = \omega(1 + i)e^{\omega x}, \qquad \ddot{W}^+ = \omega^2(1 + i)e^{\omega x}, \dots$$

and it is easy to see that the n-th (F, G)-derivative of W^+ has the form

$$W^{+[n]} = \omega^n W^+.$$

We obtain that the Taylor coefficients (in formal powers) of the function (7.17) at the origin have the simple form

$$a_n^+ = \frac{\omega^n}{n!}(1 + i). \tag{7.18}$$

In a similar way we study the case of the function u^-. The corresponding pseudoanalytic function W^- has the form $W^- = (1 - i)e^{-\omega x}$, and the corresponding Taylor coefficients at the origin are

$$a_n^- = \frac{(-\omega)^n}{n!}(1 - i). \tag{7.19}$$

Let us notice that from the fulfillment of (7.13) with the coefficients of the form (7.18) or (7.19) there follows the fulfillment of (7.14) and of all subsequent equations corresponding to $\tilde{X}^{(2)}$, $X^{(3)}$, etc. This is because of the fact that $a_{n+1}^{\pm} = \frac{\pm\omega}{n+1}a_n^{\pm}$.

7.4 Solution of the one-dimensional Schrödinger equation

Now with the aid of the sets of coefficients (7.18) and (7.19) we proceed in obtaining the general solution of (7.2) with any potential q for which a solution g_0 of (7.1) satisfying the nonzero and boundedness requirements exists. From (7.6) we have that the general solution of (7.2) has the form

$$u = c_1 u_1 + c_2 u_2$$

where c_1 and c_2 are arbitrary real constants and u_1, u_2 are defined as

$$u_1(x) = g_0(x)e^{\omega y}\sum_{n=0}^{\infty}\frac{\omega^n}{n!}\left(\text{Re}\,_*Z^{(n)}(1, 0, z) + \text{Re}\,_*Z^{(n)}(i, 0, z)\right)$$

and

$$u_2(x) = g_0(x)e^{\omega y} \sum_{n=0}^{\infty} \frac{(-\omega)^n}{n!} \left(\mathrm{Re} \,_*Z^{(n)}(1,0,z) - \mathrm{Re} \,_*Z^{(n)}(i,0,z) \right),$$

which according to (7.8) and (7.9) can be written in the form

$$u_1(x) = g_0(x)e^{\omega y} \left(\sum_{\mathrm{even}\ n=0}^{\infty} \frac{\omega^n}{n!} \left(\sum_{\mathrm{even}\ k=0}^{n} i^{n-k} \binom{n}{k} \widetilde{X}^{(k)} Y^{(n-k)} \right. \right.$$
$$+ \sum_{\mathrm{odd}\ k=1}^{n} i^{n-k+1} \binom{n}{k} X^{(k)} \widetilde{Y}^{(n-k)} \right)$$
$$+ \sum_{\mathrm{odd}\ n=1}^{\infty} \frac{\omega^n}{n!} \left(\sum_{\mathrm{odd}\ k=1}^{n} i^{n-k} \binom{n}{k} X^{(k)} Y^{(n-k)} \right.$$
$$\left. \left. + \sum_{\mathrm{even}\ k=0}^{n} i^{n-k+1} \binom{n}{k} \widetilde{X}^{(k)} \widetilde{Y}^{(n-k)} \right) \right)$$

and

$$u_2(x) = g_0(x)e^{\omega y} \left(\sum_{\mathrm{even}\ n=0}^{\infty} \frac{\omega^n}{n!} \left(\sum_{\mathrm{even}\ k=0}^{n} i^{n-k} \binom{n}{k} \widetilde{X}^{(k)} Y^{(n-k)} \right. \right.$$
$$- \sum_{\mathrm{odd}\ k=1}^{n} i^{n-k+1} \binom{n}{k} X^{(k)} \widetilde{Y}^{(n-k)} \right)$$
$$- \sum_{\mathrm{odd}\ n=1}^{\infty} \frac{\omega^n}{n!} \left(\sum_{\mathrm{odd}\ k=1}^{n} i^{n-k} \binom{n}{k} X^{(k)} Y^{(n-k)} \right.$$
$$\left. \left. - \sum_{\mathrm{even}\ k=0}^{n} i^{n-k+1} \binom{n}{k} \widetilde{X}^{(k)} \widetilde{Y}^{(n-k)} \right) \right).$$

As we know that both expressions are independent of y, in order to simplify them we can substitute any value of y. Of course, the easiest way is to substitute $y = 0$ because by definition all $Y^{(n)}(0)$ and $\widetilde{Y}^{(n)}(0)$ for $n \geq 1$ are equal to zero, and $Y^{(0)}(0) = \widetilde{Y}^{(0)}(0) = 1$. Thus, finally we obtain

$$u_1(x) = g_0(x) \left(\sum_{\mathrm{even}\ n=0}^{\infty} \frac{\omega^n}{n!} \widetilde{X}^{(n)} + \sum_{\mathrm{odd}\ n=1}^{\infty} \frac{\omega^n}{n!} X^{(n)} \right) \qquad (7.20)$$

and

$$u_2(x) = g_0(x) \left(\sum_{\mathrm{even}\ n=0}^{\infty} \frac{\omega^n}{n!} \widetilde{X}^{(n)} - \sum_{\mathrm{odd}\ n=1}^{\infty} \frac{\omega^n}{n!} X^{(n)} \right) \qquad (7.21)$$

where

$$\widetilde{X}^{(0)} \equiv 1, \quad X^{(0)} \equiv 1, \qquad (7.22)$$

$$\widetilde{X}^{(n)}(x) = \begin{cases} n \int_0^x \widetilde{X}^{(n-1)}(\xi)g_0^2(\xi)d\xi & \text{for an odd } n, \\ n \int_0^x \widetilde{X}^{(n-1)}(\xi)g_0^{-2}(\xi)d\xi & \text{for an even } n, \end{cases} \tag{7.23}$$

$$X^{(n)}(x) = \begin{cases} n \int_0^x X^{(n-1)}(\xi)g_0^{-2}(\xi)d\xi & \text{for an odd } n, \\ n \int_0^x X^{(n-1)}(\xi)g_0^2(\xi)d\xi & \text{for an even } n. \end{cases} \tag{7.24}$$

In the next subsection we validate this result by a direct substitution into equation (7.2).

7.5 Validating the result

In order to substitute (7.20) and (7.21) or equivalently

$$v_1(x) = g_0(x) \sum_{\text{even } n=0}^{\infty} \frac{\omega^n}{n!} \widetilde{X}^{(n)} \quad \text{and} \quad v_2(x) = g_0(x) \sum_{\text{odd } n=1}^{\infty} \frac{\omega^n}{n!} X^{(n)}$$

into equation (7.2) we first make some helpful observations.

It is well known that a nonvanishing solution g_0 of (7.1) allows us to factorize the Schrödinger operator as

$$\partial_x^2 - q(x) = \left(\partial_x + \frac{g_0'}{g_0}\right)\left(\partial_x - \frac{g_0'}{g_0}\right). \tag{7.25}$$

The first-order operators in their turn can be factorized as well, so we obtain

$$\partial_x^2 - q = g_0^{-1}\partial_x g_0^2 \partial_x g_0^{-1}.$$

Now let us consider v_1. By definition, for an even n we have

$$\widetilde{X}^{(n)}(x) = n \int_0^x \widetilde{X}^{(n-1)}(\xi)\frac{d\xi}{g_0^2(\xi)}.$$

Thus, application of the operator $\partial_x^2 - q$ to $g_0\widetilde{X}^{(n)}$ for an even n and $n \geq 2$ (for $n = 0$ the result is zero) gives us

$$\left(\partial_x^2 - q\right)\left(g_0\widetilde{X}^{(n)}\right) = g_0^{-1}\partial_x g_0^2 \partial_x \widetilde{X}^{(n)} = ng_0^{-1}\partial_x \widetilde{X}^{(n-1)}$$
$$= (n-1)ng_0\widetilde{X}^{(n-2)}.$$

Then

$$\left(\partial_x^2 - q\right) v_1 = g_0 \sum_{\text{even } n=2}^{\infty} \frac{\omega^n}{(n-2)!} \widetilde{X}^{(n-2)}$$

$$= \omega^2 g_0 \sum_{\text{even } n=0}^{\infty} \frac{\omega^n}{n!} \widetilde{X}^{(n)} = \omega^2 v_1.$$

In a similar way one can verify that v_2 is a solution of (7.2) as well. Note that according to the general result formulated in Subsection 4.2, both series in v_1 and v_2 are uniformly convergent on the interval I_x. This fact can be quite easily verified as well by estimating the integrals in $\widetilde{X}^{(n)}$ and $X^{(n)}$ by the supremum of the functions g_0^2 and g_0^{-2} multiplied by successive antiderivatives of x.

7.6   The "−" case

Consider the equation

$$(-\partial_x^2 + q(x) - \omega^2)u(x) = 0 \tag{7.26}$$

and the corresponding two-dimensional equation

$$(-\Delta + q(x) - \omega^2)U(x,y) = 0. \tag{7.27}$$

Its particular solution can be chosen as

$$f(x,y) = g_0(x) \cos \omega y$$

which is different from zero on the rectangular domain $\Omega = I_x \times (-\frac{\pi}{2\omega}, \frac{\pi}{2\omega})$. In order to obtain the general solution of (7.26) in fact we should only obtain two sets of Taylor coefficients as in Subsection 7.3. For this, once more we take $q \equiv 0$ and consider two linearly independent solutions of the equation $(\partial_x^2 + \omega^2)u(x) = 0$, $u^+(x) = \cos \omega x$ and $u^-(x) = \sin \omega x$. The next step is to construct v^+ and v^-. We have

$$v^+ = \frac{1}{\cos \omega y} \overline{A} \left(i \cos^2 \omega y \partial_{\overline{z}} \left(\frac{\cos \omega x}{\cos \omega y} \right) \right) = - \sin \omega x \tan \omega y$$

(we have fixed the arbitrary constant as zero). Thus,

$$W^+ = \cos \omega x - i \sin \omega x \tan \omega y.$$

In a similar way we obtain

$$W^- = \sin \omega x + i \cos \omega x \tan \omega y.$$

Noting that the definition of the (F, G)-derivative in this case has the form

$$\dot{W} = W_z - \frac{i\omega}{2} \tan \omega y \overline{W}$$

we obtain the relations $\dot{W}^+ = -\omega W^-$ and $\dot{W}^- = \omega W^+$ and hence the following formulas for the corresponding Taylor coefficients in formal powers in the origin

$$a_n^+ = \frac{(i\omega)^n}{n!} \text{ for an even } n \quad \text{and} \quad a_n^+ = 0 \text{ for an odd } n,$$

$$a_n^- = 0 \text{ for an even } n \quad \text{and} \quad a_n^- = \frac{-i(i\omega)^n}{n!} \text{ for an odd } n.$$

Thus we arrive at the following general solution of equation (7.26) for any potential q admitting a particular solution g_0 with the described above properties:

$$u = c_1 u_1 + c_2 u_2$$

with

$$u_1(x) = g_0(x) \sum_{\text{even } n=0}^{\infty} \frac{(i\omega)^n}{n!} \widetilde{X}^{(n)}$$

and

$$u_2(x) = g_0(x) \sum_{\text{odd } n=1}^{\infty} \frac{i(i\omega)^n}{n!} X^{(n)}$$

where $X^{(n)}$ and $\widetilde{X}^{(n)}$ are defined by (7.22)–(7.24).

7.7 Complex potential

It is clear that the results obtained in the preceding subsections remain valid in the case of a complex-valued potential q and an arbitrary complex number ω. Consider the equation

$$(-\partial_x^2 + q(x) + \omega^2)u(x) = 0 \tag{7.28}$$

where q and u are complex-valued and ω is any complex number. We assume that g_0 is a nonvanishing solution of the equation $(-\partial_x^2 + q(x))g_0 = 0$ satisfying the boundedness requirements. Then the general solution of (7.28) has the form

$$u = c_1 u_1 + c_2 u_2 \tag{7.29}$$

where c_1 and c_2 are arbitrary complex constants and u_1, u_2 are defined as

$$u_1 = g_0 \sum_{\text{even } n=0}^{\infty} \frac{\omega^n}{n!} \widetilde{X}^{(n)} \tag{7.30}$$

and

$$u_2 = g_0 \sum_{\text{odd } n=1}^{\infty} \frac{\omega^n}{n!} X^{(n)} \tag{7.31}$$

where as before $X^{(n)}$ and $\widetilde{X}^{(n)}$ are defined by (7.22)–(7.24).

7.8 Solution of the Sturm-Liouville equation

Having obtained the general solution of (7.28) it is easy to obtain the general solution of the more general Sturm-Liouville equation

$$\partial_x(p\partial_x u) + qu = \omega^2 u \tag{7.32}$$

where $p \in C^2(I_x)$ is a nonvanishing complex-valued function, q and u satisfy conditions from the preceding subsection. We assume that

$$\partial_x(p\partial_x g_0) + qg_0 = 0 \tag{7.33}$$

and observe that the following factorization holds:

$$(\partial_x p\partial_x + q)u = p^{1/2}\left(\partial_x + \frac{g'}{g}\right)\left(\partial_x - \frac{g'}{g}\right)(p^{1/2}u) = g_0^{-1}\partial_x\left(g^2\partial_x\left(g_0^{-1}u\right)\right)$$

where $g = p^{1/2}g_0$ and by analogy with (7.29)–(7.31) we obtain the solution of (7.32),

$$u = c_1 u_1 + c_2 u_2 \tag{7.34}$$

where

$$u_1 = g_0 \sum_{\text{even } n=0}^{\infty} \frac{\omega^n}{n!}\widetilde{X}^{(n)} \tag{7.35}$$

and

$$u_2 = g_0 \sum_{\text{odd } n=1}^{\infty} \frac{\omega^{n-1}}{n!}X^{(n)} \tag{7.36}$$

where the definition of $X^{(n)}$ and $\widetilde{X}^{(n)}$ is slightly modified:

$$\widetilde{X}^{(0)} \equiv 1, \qquad X^{(0)} \equiv 1, \tag{7.37}$$

$$\widetilde{X}^{(n)}(x) = \begin{cases} n\int_0^x \widetilde{X}^{(n-1)}(\xi)g_0^2(\xi)d\xi & \text{for an odd } n, \\ n\int_0^x \widetilde{X}^{(n-1)}(\xi)g^{-2}(\xi)d\xi & \text{for an even } n, \end{cases} \tag{7.38}$$

$$X^{(n)}(x) = \begin{cases} n\int_0^x X^{(n-1)}(\xi)g^{-2}(\xi)d\xi & \text{for an odd } n, \\ n\int_0^x X^{(n-1)}(\xi)g_0^2(\xi)d\xi & \text{for an even } n. \end{cases} \tag{7.39}$$

Let us verify that u_1 is indeed a solution of (7.32). We have

$$(\partial_x p \partial_x + q)u_1 = g_0^{-1} \partial_x \left(g^2 \partial_x \sum_{\text{even } n=0}^{\infty} \frac{\omega^n}{n!} \widetilde{X}^{(n)} \right)$$

$$= g_0^{-1} \sum_{\text{even } n=2}^{\infty} \frac{\omega^n}{(n-1)!} \partial_x \widetilde{X}^{(n-1)}$$

$$= \omega^2 g_0 \sum_{\text{even } n=2}^{\infty} \frac{\omega^{n-2}}{(n-2)!} \widetilde{X}^{(n-2)} = \omega^2 u_1.$$

In a similar way the solution u_2 can be verified as well.

Thus, we have proved the following theorem.

Theorem 95 ([74]). *Let p and q be complex-valued functions of an independent real variable $x \in [0, a]$, $p \in C^1(0, a)$ be bounded and nonvanishing on $[0, a]$ and ω be an arbitrary complex number. Suppose that there exists a solution g_0 of the equation $(pg_0')' + qg_0 = 0$ on $(0, a)$ such that $g_0 \in C^2(0, a)$ together with $1/g_0$ are bounded on $[0, a]$. Then the general solution of (7.32) has the form*

$$u = c_1 u_1 + c_2 u_2$$

where c_1 and c_2 are arbitrary complex constants,

$$u_1 = g_0 \sum_{\text{even } n=0}^{\infty} \frac{\omega^n}{n!} \widetilde{X}^{(n)} \quad and \quad u_2 = g_0 \sum_{\text{odd } n=1}^{\infty} \frac{\omega^{n-1}}{n!} X^{(n)}$$

with $\widetilde{X}^{(n)}$ and $X^{(n)}$ being defined by (7.37)–(7.39) and both series converge uniformly on $[0, a]$.

We obtained this theorem under too restrictive conditions. Here we formulate its generalization as was done in [78], which allows us to obtain a general solution of the Sturm-Liouville equation

$$(pu')' + qu = \lambda r u \tag{7.40}$$

in the form of a spectral parameter power series. We give a proof of this result which does not depend on pseudoanalytic function theory.

Theorem 96 ([78]). *Assume that on a finite interval $[a, b]$, equation*

$$(pv')' + qv = 0, \tag{7.41}$$

possesses a particular solution u_0 such that the functions $u_0^2 r$ and $1/(u_0^2 p)$ are continuous on $[a, b]$. Then the general solution of (7.40) on (a, b) has the form

$$u = c_1 u_1 + c_2 u_2 \tag{7.42}$$

where c_1 and c_2 are arbitrary complex constants,

$$u_1 = u_0 \sum_{k=0}^{\infty} \lambda^k \widetilde{X}^{(2k)} \quad and \quad u_2 = u_0 \sum_{k=0}^{\infty} \lambda^k X^{(2k+1)} \tag{7.43}$$

with $\widetilde{X}^{(n)}$ and $X^{(n)}$ being defined by the recursive relations

$$\widetilde{X}^{(0)} \equiv 1, \qquad X^{(0)} \equiv 1, \tag{7.44}$$

$$\widetilde{X}^{(n)}(x) = \begin{cases} \int\limits_{x_0}^{x} \widetilde{X}^{(n-1)}(s) u_0^2(s) r(s)\, ds, & n \text{ odd}, \\ \int\limits_{x_0}^{x} \widetilde{X}^{(n-1)}(s) \frac{1}{u_0^2(s) p(s)}\, ds, & n \text{ even}, \end{cases} \tag{7.45}$$

$$X^{(n)}(x) = \begin{cases} \int\limits_{x_0}^{x} X^{(n-1)}(s) \frac{1}{u_0^2(s) p(s)}\, ds, & n \text{ odd}, \\ \int\limits_{x_0}^{x} X^{(n-1)}(s) u_0^2(s) r(s)\, ds, & n \text{ even}, \end{cases} \tag{7.46}$$

where x_0 is an arbitrary point in $[a,b]$ such that p is continuous at x_0 and $p(x_0) \neq 0$. Further, both series in (7.43) converge uniformly on $[a,b]$.

Proof. First we prove that u_1 and u_2 are indeed solutions of (7.40) whenever the application of the operator $L = \frac{d}{dx} p \frac{d}{dx} + q$ to them makes sense. For this, note that if $Lu_0 = 0$, then L can be written in the factorized form $L = \frac{1}{u_0} \frac{d}{dx} p\, u_0^2 \frac{d}{dx} \frac{1}{u_0}$. Then application of $\frac{1}{r} L$ to u_1 gives

$$\frac{1}{r} L u_1 = \frac{1}{r u_0} \frac{d}{dx} \left(p u_0^2 \frac{d}{dx} \sum_{k=0}^{\infty} \lambda^k \widetilde{X}^{(2k)} \right) = \frac{1}{r u_0} \frac{d}{dx} \sum_{k=1}^{\infty} \lambda^k \widetilde{X}^{(2k-1)}$$

$$= u_0 \sum_{k=1}^{\infty} \lambda^k \widetilde{X}^{(2k-2)} = \lambda u_1.$$

In a similar way one can check that u_2 satisfies (7.40) as well. In order to give sense to this chain of equalities it is sufficient to prove the uniform convergence of the series involved in u_1 and u_2 as well as of the series obtained by a term-wise differentiation. This can be done with the aid of the Weierstrass M-test. Indeed, we have $\left| \widetilde{X}^{(2k)} \right| \leq \left(\max \left| r u_0^2 \right| \right)^k \left(\max \left| \frac{1}{p u_0^2} \right| \right)^k \frac{|b-a|^{2k}}{(2k)!}$ and the series $\sum_{k=0}^{\infty} \frac{c^k}{(2k)!}$ is convergent where

$$c = |\lambda| \left(\max \left| r u_0^2 \right| \right) \left(\max \left| \frac{1}{p u_0^2} \right| \right) |b-a|^2 . \tag{7.47}$$

The uniform convergence of the series in u_2 as well as of the series of derivatives can be shown similarly.

The last step is to verify that the Wronskian of u_1 and u_2 is different from zero at least at one point (which necessarily implies the linear independence of u_1 and u_2 on the whole segment $[a,b]$). It is easy to see that by definition all the $\widetilde{X}^{(n)}(x_0)$ and $X^{(n)}(x_0)$ vanish except for $\widetilde{X}^{(0)}(x_0)$ and $X^{(0)}(x_0)$ which equal 1. Thus

$$u_1(x_0) = u_0(x_0), \qquad u_1'(x_0) = u_0'(x_0), \tag{7.48}$$

$$u_2(x_0) = 0, \qquad u_2'(x_0) = \frac{1}{u_0(x_0)p(x_0)} \tag{7.49}$$

and the Wronskian of u_1 and u_2 at x_0 equals $1/p(x_0) \neq 0$. $\qquad\square$

Remark 97. In the case $\lambda = 0$, the solution (7.43) becomes $u_1 = u_0$ and $u_2 = u_0 \int_{x_0}^{x} \frac{ds}{u_0^2(s)p(s)}$. The expression for u_2 is a well-known formula for constructing a second linearly independent solution.

Remark 98. The result of Theorem 96 is valid for infinite intervals as well, the series being uniformly convergent on any finite subinterval.

Remark 99. One of the functions ru_0^2 or $1/(pu_0^2)$ may not be continuous on $[a,b]$ and yet u_1 or u_2 may make sense. For example, in the case of the Bessel equation $(xu')' - \frac{1}{x}u = -\lambda xu$, we can choose $u_0(x) = x/2$. Then $1/(pu_0^2) \notin C[0,1]$. Nevertheless all integrals in (7.45) exist and u_1 coincides with the nonsingular $J_1(\sqrt{\lambda}x)$, while u_2 is a singular solution of the Bessel equation.

Remark 100. In the regular case the existence and construction of the required u_0 presents no difficulty. Let p and q be real-valued, $p(x) \neq 0$ for all $x \in [a,b]$ and let p, p', r and q be continuous on $[a,b]$. Then (7.41) possesses two linearly independent regular solutions v_1 and v_2 whose zeros alternate. Thus one may choose $u_0 = v_1 + iv_2$.

Remark 101. The procedure for construction of solutions described in Theorem 96 works not only when a solution is available for $\lambda = 0$, but in fact when a solution of the equation

$$(pu_0')' + qu_0 = \lambda_0 ru_0 \tag{7.50}$$

is known for some fixed λ_0. The solution (7.43) now takes the form

$$u_1 = u_0 \sum_{k=0}^{\infty} (\lambda - \lambda_0)^k \widetilde{X}^{(2k)} \quad \text{and} \quad u_2 = u_0 \sum_{k=0}^{\infty} (\lambda - \lambda_0)^k X^{(2k+1)}.$$

This can be easily verified by writing (7.40) as

$$(L - \lambda_0 r)u = (\lambda - \lambda_0)ru.$$

The operator on the left-hand side can be factorized exactly as in the proof of the theorem, and the same reasoning carries through.

Remark 102. For calculating the series in (7.43) it may be convenient to calculate $X^{(n)}$ or $\widetilde{X}^{(n)}$ directly from $X^{(n-2)}$ or $\widetilde{X}^{(n-2)}$. For example, when n is even we have

$$\widetilde{X}^{(n)}(x) = \int_{x_0}^{x} \frac{1}{u_0(s)^2 p(s)} \int_{x_0}^{s} u_0(t)^2 r(t) \widetilde{X}^{(n-2)}(t)\, dt\, ds$$

$$= \int_{x_0}^{x} (P(x) - P(t)) u_0(t)^2 r(t) \widetilde{X}^{(n-2)}(t)\, dt$$

where $P' = 1/(u_0^2 p)$.

Remark 103. Other representations of the general solution of (7.40) as a formal power series have been long known (see [104, Theorem 1], [25]) and used for studying qualitative properties of solutions. The complicated manner in which the parameter λ appears in those representations makes that form of a general solution too difficult for quantitative analysis of spectral and boundary value problems. In contrast, the solution (7.42)–(7.46) is a power series with respect to λ, making it quite attractive for numerical solution of spectral, initial value and boundary value problems.

A special case of Theorem 96, with $q \equiv 0$, $\lambda = 1$, was known to H. Weyl.

Corollary 104 ([124]). *Let $1/p$ and r be continuous on $[a, b]$. The general solution of the equation*

$$(pu')' = ru \tag{7.51}$$

on (a, b) has the form

$$u = c_1 u_1 + c_2 u_2 \tag{7.52}$$

where c_1 and c_2 are arbitrary constants and u_1, u_2 are defined by (7.43)–(7.46) with $u_0 \equiv \lambda = 1$.

This corollary enables us to find the particular solution u_0 discussed in Remark 100.

7.9 Numerical method for solving Sturm-Liouville equations

In this section following the exposition from [78] we consider the Sturm-Liouville equation (7.40) on $[a, b]$ with any desired initial conditions. The numerical implementation of the solution via the representation (7.43) for a general solution is algorithmically simple. One must consider the accuracy of calculation of the iterated integrals in (7.45) and (7.46), and the rate of convergence of the series (7.43), because in numerical work one must work with finitely many terms.

The main parameters that one can control are the number M of subintervals in which to divide $[a, b]$ when integrating numerically and the number N of powers in the truncated series. The relationship of M to the accuracy of the integrals is a standard question and will not be discussed here. In regards to N, observe that one can not always expect a good approximation to u over all of $[a, b]$ with a series of N terms, no matter how precisely the integrals are calculated. However, using the estimate for $\left|\widetilde{X}^{(2k)}\right|$ and $\left|X^{(2k-1)}\right|$ (see the proof of Theorem 96 below) it is easy to obtain a rough but useful estimate for the tail of the series. Namely, consider $|u_1 - u_{1,N}|$ where $u_{1,N} = u_0 \sum\limits_{k=0}^{N} \lambda^k \widetilde{X}^{(2k)}$. We have

$$|u_1 - u_{1,N}| = |u_0| \left| \sum_{k=N+1}^{\infty} \lambda^k \widetilde{X}^{(2k)} \right| \leq \max |u_0| \sum_{k=N+1}^{\infty} \frac{c^k}{(2k)!}$$

$$= \max |u_0| \left| \cosh \sqrt{c} - \sum_{k=0}^{N} \frac{c^k}{(2k)!} \right|$$

where c is defined by (7.47). In a similar way one can see that the remainder of the series corresponding to u_2 is estimated by the tail of the power series of $\sinh \sqrt{c}$. Thus, if a certain value of N is seen to be insufficient for achieving a required accuracy, the interval can be subdivided and the initial value problem solved on the first subinterval. The initial values of the solution for the second subinterval are calculated easily taking into account that $u_1' = \frac{u_0'}{u_0} u_1 + \frac{1}{u_0 p} \sum\limits_{k=1}^{\infty} \lambda^k \widetilde{X}^{(2k-1)}$ (and analogously for u_2). Thus, no numerical differentiation is necessary and this process can be continued with little loss in accuracy.

The required particular solution u_0 may be calculated using any available algorithm; in the examples presented below we have applied the formula of Corollary 104, applying the above subdivision procedure. All of the calculations were performed with *Mathematica* (Wolfram).

Chapter 8

Spectral Problems and Darboux Transformation

8.1 Sturm-Liouville problem as a problem of finding zeros of an analytic function

The fact that spectral Sturm-Liouville problems are related to the problem of finding zeros of complex analytic functions of the variable λ is quite well known (see, e.g., [86]). For a regular Sturm-Liouville problem the corresponding analytic function is even entire. The representation (7.42)–(7.46) allows us to obtain the Taylor series of that analytic function explicitly. As an example, let us first consider a spectral problem for (7.40) with the boundary conditions

$$u(0) = 0 \quad \text{and} \quad u(1) = 0.$$

We suppose that the coefficients satisfy the conditions from Remark 100 and that u_0 is constructed as described there, taking $x_0 = 0$. From the first boundary condition and (7.48), the constant c_1 in (7.42) must be zero. Then the spectral problem reduces to finding values of λ for which $u_2(1) = u_0(1) \sum\limits_{k=0}^{\infty} \lambda^k X^{(2k+1)}(1)$ vanishes. In other words, this spectral problem reduces to the calculation of zeros of the complex analytic function $\kappa(\lambda) = \sum\limits_{m=0}^{\infty} a_m \lambda^m$ where

$$a_m = u_0(1) X^{(2k+1)}(1).$$

Now let α and β be arbitrary real numbers and consider the more general boundary conditions

$$u(a)\cos\alpha + u'(a)\sin\alpha = 0, \tag{8.1}$$
$$u(b)\cos\beta + u'(b)\sin\beta = 0 \tag{8.2}$$

together with equation (7.40). Taking the solutions u_1 and u_2 defined by (7.43) and using (7.48), (7.49) with $x_0 = a$, we obtain from (8.1) the equation

$$c_1(u_0(a)\cos\alpha + u_0'(a)\sin\alpha) + c_2\frac{\sin\alpha}{u_0(a)p(a)} = 0,$$

which gives $c_2 = \gamma c_1$ when $\alpha \neq \pi n$, with $\gamma = -u_0(a)p(a)(u_0(a)\cot\alpha + u_0'(a))$, whereas $c_1 = 0$ when $\alpha = \pi n$. In the latter case the result is similar to the example considered above, thus let us suppose $\alpha \neq \pi n$. From the definition of u_1 and u_2 we have

$$u_1' = \frac{u_0'}{u_0}u_1 + \frac{1}{u_0 p}\sum_{k=1}^{\infty}\lambda^k \widetilde{X}^{(2k-1)} \quad \text{and} \quad u_2' = \frac{u_0'}{u_0}u_2 + \frac{1}{u_0 p}\sum_{k=0}^{\infty}\lambda^k X^{(2k)}.$$

Then the boundary condition (8.2) implies that

$$(u_0(b)\cos\beta + u_0'(b)\sin\beta)\left(\sum_{k=0}^{\infty}\lambda^k \widetilde{X}^{(2k)}(b) + \gamma\sum_{k=0}^{\infty}\lambda^k X^{(2k+1)}(b)\right)$$

$$+\frac{\sin\beta}{u_0(b)p(b)}\left(\sum_{k=1}^{\infty}\lambda^k \widetilde{X}^{(2k-1)}(b) + \gamma\sum_{k=0}^{\infty}\lambda^k X^{(2k)}(b)\right) = 0.$$

Thus the spectral problem (7.40), (8.1), (8.2) reduces to the problem of calculating zeros of the analytic function $\kappa(\lambda) = \sum_{m=0}^{\infty}a_m\lambda^m$ where

$$a_0 = (u_0(b)\cos\beta + u_0'(b)\sin\beta)(1 + \gamma X^{(1)}(b)) + \frac{\gamma\sin\beta}{u_0(b)p(b)}$$

and

$$a_m = (u_0(b)\cos\beta + u_0'(b)\sin\beta)\left(\widetilde{X}^{(2m)}(b) + \gamma X^{(2m+1)}(b)\right)$$

$$+\frac{\sin\beta}{u_0(b)p(b)}\left(\widetilde{X}^{(2m-1)}(b) + \gamma X^{(2m)}(b)\right), \quad m = 1, 2, \ldots.$$

This reduction of a Sturm-Liouville spectral problem lends itself to a simple numerical implementation. To calculate the first n eigenvalues we consider the Taylor polynomial $\kappa_N(\lambda) = \sum_{m=0}^{N}a_m\lambda^m$ with $N \geq n$. Thus the numerical approximation of eigenvalues of the Sturm-Liouville problem reduces to the calculation of zeros of the polynomial $\kappa_N(\lambda)$.

There is no need to work with zeros of only one polynomial. It is well known that in general the higher roots of a polynomial become less stable with respect to small inaccuracies in coefficients. Our spectral parameter power series (SPPS) method (the term introduced in [78]) is well suited to overcome this problem and thus to calculate higher eigenvalues with a good accuracy. This is done using

Remark 101. Suppose we have already calculated the eigenvalue λ_0 using the procedure described above as a first root of the obtained polynomial. Then for the next step we define $U_0 = u_1 + iu_2$ where u_1 and u_2 are defined by (7.43) with $\lambda = \lambda_0$. The function U_0 is then a solution of (7.50). We use it to obtain the eigenvalue λ_1 of the original problem observing that $\lambda_1 = \Lambda_1 + \lambda_0$ where Λ_1 is the first eigenvalue of the equation $(L - \lambda_0 r)u = \Lambda u$ with the same boundary conditions as in the original problem. This procedure can be continued for calculating higher eigenvalues. Note that if $\lambda_0 = 0$ we should begin this shifting procedure starting with λ_1.

8.1.1 Sturm-Liouville problems with spectral parameter dependent boundary conditions

In this subsection based on [78] we consider Sturm-Liouville problems of the form

$$(pu')' + qu = \lambda r u, \quad x \in [a, b], \tag{8.3}$$

$$u(a)\cos\alpha + u'(a)\sin\alpha = 0, \qquad \alpha \in [0, \pi), \tag{8.4}$$

$$\beta_1 u(b) - \beta_2 u'(b) = \varphi(\lambda)\left(\beta_1' u(b) - \beta_2' u'(b)\right), \tag{8.5}$$

where φ is a complex-valued function of the variable λ and β_1, β_2, β_1', β_2' are complex numbers. This kind of problem arises in many physical applications (we refer to [7] and references therein) and has been studied in a considerable number of publications [7, 26, 29, 31, 45, 121]. For some special forms of the function φ such as $\varphi(\lambda) = \lambda$ or $\varphi(\lambda) = \lambda^2 + c_1\lambda + c_2$, results were obtained [29], [121] concerning the regularity of the problem (8.3)–(8.5); we will not dwell upon the details. Our purpose is to show the applicability of the spectral parameter power series (SPPS) method to this type of Sturm-Liouville problems. For simplicity, let us suppose that $\alpha = 0$ and hence the condition (8.4) becomes $u(a) = 0$. Then as was shown in the preceding section, if an eigenfunction exists it necessarily coincides with u_2 up to a multiplicative constant.

In this case condition (8.5) becomes equivalent to the equality

$$\left(u_0(b)\varphi_1(\lambda) - u_0'(b)\varphi_2(\lambda)\right)\sum_{k=0}^{\infty}\lambda^k X^{(2k+1)}(b) - \frac{\varphi_2(\lambda)}{u_0(b)p(b)}\sum_{k=0}^{\infty}\lambda^k X^{(2k)}(b) = 0 \tag{8.6}$$

where $\varphi_{1,2}(\lambda) = \beta_{1,2} - \beta_{1,2}'\varphi(\lambda)$. Calculation of eigenvalues given by (8.6) is especially simple in the case of φ being a polynomial of λ. Precisely this particular situation was considered in all of the above-mentioned references concerning Sturm-Liouville problems with spectral parameter dependent boundary conditions. For these problems the calculation of eigenvalues using our method does not present any additional difficulty compared to the parameter independent situation discussed in the preceding section.

8.2   Numerical method for solving Sturm-Liouville problems

Here we discuss some numerical examples of application of the SPPS method based on the result of Theorem 96.

Paine Problem. A number of spectral problems which have become standard test cases appear in [98, 106]. As a first example we consider

$$p(x) = -1, \qquad q(x) = \frac{1}{(x+0.1)^2},$$

$$u(0) = 0, \qquad u(\pi) = 0.$$

The eigenvalues in the following table were calculated via SPPS using integration on 10,000 subintervals for calculating $N = 100$ powers of λ. These eigenvalues were found as roots of a single polynomial (i.e., the shifting of λ as described in Remark 101 was not applied). Due to the sensitivity of the larger roots of the polynomial to errors in the coefficients, 100-digit arithmetic was used.

n	λ_n [106]	λ_n SPPS [78]
0	1.5198658211	1.519865821099
1	4.9433098221	4.943309822144
2	10.284662645	10.28466264509
3	17.559957746	17.55995774633
4	26.782863158	26.78286315899
5	37.964425862	37.96442587941
6	51.113357757	51.11335707578
7	66.236447704	66.23646092491
8	83.338962374	83.33879073183
9	102.42498840	102.4259718823
10	123.49770680	123.512483827

On the basis of the above values, a new calculation was made by shifting with $\lambda^* = 66$, resulting in the following improved approximations for the last few eigenvalues.

n	λ_n [106]	λ_n SPPS [78]
7	66.236447704	66.23644770359
8	83.338962374	83.33896237419
9	102.42498840	102.42498839828
10	123.49770680	123.49770680101
11	146.55960608	146.55960605783
12	171.61264485	171.61265439928

With $\lambda^* = 146$ and increasing the number of powers to $N = 150$, the following further values were obtained.

n	λ_n [106]	λ_n SPPS [78]
11	146.55960608	146.55586199495330
12	171.61264485	171.60875781110985
13	198.65837500	198.65416389844202

When the number of digits for internal calculations was increased to 150, SPPS produced the same results.

Coffey-Evans equation. This test case, defined by

$$p(x) = -1, \qquad q(x) = -2\beta \cos 2x + \beta^2 \sin^2 2x,$$
$$u(-\pi/2) = 0, \qquad u(\pi/2) = 0$$

presents the challenge of distinguishing eigenvalues within the triple clusters which form as the parameter β increases. We present results for $\beta = 20, 30, 50$. In all cases given here the eigenvalues were obtained without shifting λ.

$\beta = 20$. $M = 10,000$ subintervals, $N = 180$ powers, 100 digits of precision.

n	λ_n [28, 85]	λ_n SPPS [78]
0	-0.00000000000000	0.0000000000000003
1	77.91619567714397	77.9161956771439703
2	151.46277834645663	151.4627783464566396
3	151.46322365765863	151.4632236576586490
4	151.46366898835165	151.4636689883516575
5	220.15422983525995	220.1542298352599497
6	283.0948	283.0948146954014377
7	283.2507	283.2507437431126800
8	283.4087	283.4087354034293064

$\beta = 30$ $M = 10,000$ subintervals, $N = 150$ powers, 100 digits of precision.

n	λ_n [85, 106]	λ_n SPPS [78]
0	0.00000000000000	0.000000000000000002
1	117.946307662070	117.94630766206876
2		231.664928928423790
3	231.66492931296	231.664928928423791
4		231.664930082035462
5		340.888299091685489
6		403.219684016171863
7		403.219684016171917

$\beta = 50$ $M = 10,000$ subintervals, $N = 150$ powers, 100 digits of precision.

n	λ_n [106]	λ_n SPPS [78]
0	0.00000000000000	0.000000000000000003
1	197.968726516507	197.96872651650729
2		391.807
3	391.80819148905	391.810
4		547.1397060

8.3   A remark on the Darboux transformation

The Darboux transformation is a very useful and important tool studied in dozens of works (see, e.g., [90]). It is closely related to the factorization of the Schrödinger operator (7.25). Consider the equation

$$\left(\partial_x + \frac{g_0'}{g_0}\right)\left(\partial_x - \frac{g_0'}{g_0}\right)u = \omega^2 u.$$

Applying the operator $\left(\partial_x - \frac{g_0'}{g_0}\right)$ to both sides and writing $v = \left(\partial_x - \frac{g_0'}{g_0}\right)u$, one obtains that solutions of equation (7.28) are transformed into solutions of another Schrödinger equation

$$\left(\partial_x - \frac{g_0'}{g_0}\right)\left(\partial_x + \frac{g_0'}{g_0}\right)v = \omega^2 v$$

which can be written also as

$$\left(-\partial_x^2 + r(x) + \omega^2\right)v(x) = 0, \tag{8.7}$$

where $r = 2\left(\frac{g_0'}{g_0}\right)^2 - q$. Now, as we are able to construct the general solution of (7.28) by a known solution of (7.1) we can also obtain an explicit form of the result of the Darboux transformation. First, let us apply the operator $\left(\partial_x - \frac{g_0'}{g_0}\right) = g_0\partial_x g_0^{-1}$ to u_1 defined by (7.30). We have

$$v_1 = \left(\partial_x - \frac{g_0'}{g_0}\right)u_1 = g_0 \sum_{\text{even } n=0}^{\infty} \frac{\omega^n}{n!}\partial_x \widetilde{X}^{(n)}$$

$$= g_0^{-1} \sum_{\text{even } n=2}^{\infty} \frac{\omega^n}{(n-1)!}\widetilde{X}^{(n-1)} = \frac{\omega}{g_0} \sum_{\text{odd } n=1}^{\infty} \frac{\omega^n}{n!}\widetilde{X}^{(n)}$$

and in a similar way we obtain

$$v_2 = \left(\partial_x - \frac{g_0'}{g_0}\right)u_2 = \frac{\omega}{g_0} \sum_{\text{even } n=0}^{\infty} \frac{\omega^n}{n!}X^{(n)}.$$

Thus, the general solution of the Schrödinger equation (8.7) obtained from (7.28) by the Darboux transformation has the form

$$v = \frac{c_1}{g_0} \sum_{\text{even } n=0}^{\infty} \frac{\omega^n}{n!} X^{(n)} + \frac{c_2}{g_0} \sum_{\text{odd } n=1}^{\infty} \frac{\omega^n}{n!} \widetilde{X}^{(n)}$$

where $X^{(n)}$ and $\widetilde{X}^{(n)}$ are defined by (7.22)–(7.24).

Part III

Applications to
Real First-order Systems

Chapter 9

Beltrami Fields

9.1   Description of the result

Solutions of the equation

$$\operatorname{rot} \overrightarrow{B} + \alpha \overrightarrow{B} = 0 \qquad (9.1)$$

where α is a scalar function of space coordinates are known as Beltrami fields and are of fundamental importance in different branches of modern physics (see, e.g., [128], [82], [43], [125], [4], [55], [50], [67]). For simplicity, here we consider the real-valued proportionality factor α and real-valued solutions of (9.1), though the presented approach is applicable in a complex-valued situation as well (instead of complex Vekua equations their bicomplex generalizations should be considered, see Section 14.3). We consider equation (9.1) on a plane of the variables x and y, that is α and $\overrightarrow{B}$ are functions of two Cartesian variables only. In this case, as we show in Section 9.2, equation (9.1) reduces to the equation

$$\operatorname{div}\left(\frac{1}{\alpha}\nabla u\right) + \alpha u = 0. \qquad (9.2)$$

This second-order equation can be reduced to a corresponding main Vekua equation. This reduction under quite general conditions allows us to construct a complete system of exact solutions of (9.2) explicitly. For the reduction of (9.2) to a Vekua equation it is sufficient to find a particular solution of (9.2). Here (Section 9.3) we show that in a (very important for applications) case of α being a function of one Cartesian variable, a particular solution of (9.2) is always available in a simple explicit form. This situation corresponds to models describing waves propagating in stratified media (see, e.g., [75]). As a result, in this case we are able to construct a complete system of solutions explicitly which for many purposes means a general solution. We give an example of such construction.

We show that when $\alpha = \alpha(y)$ (of course in a similar way the case $\alpha = \alpha(x)$ can be considered) equation (9.2) and hence equation (9.1) reduce to the main

Vekua equation of the form

$$\partial_{\bar{z}} W(x, y) = \frac{if'(y)}{2f(y)} \overline{W}(x, y) \qquad (9.3)$$

where

$$f = \frac{c_1}{\sqrt{\alpha}} \sin \mathcal{A} + \frac{c_2}{\sqrt{\alpha}} \cos \mathcal{A};$$

$\mathcal{A}$ is an antiderivative of α with respect to y, c_1 and c_2 are arbitrary real constants. A complete (in a compact uniform convergence topology) system of exact solutions to (9.3) can be constructed explicitly. Thus, in the case when α is a function of one Cartesian variable, the Vekua equation equivalent to (9.1) can be solved and a complete system of solutions of (9.1) is obtained.

The results of this chapter have been obtained in a joint work with H. Oviedo [77].

9.2 Reduction of (9.1) to a Vekua equation

We consider equation (9.1) where both α and $\vec{B}$ are supposed to be dependent on two Cartesian variables x and y. Then equation (9.1) can be written as the system

$$\partial_y B_3 + \alpha B_1 = 0, \qquad (9.4)$$
$$-\partial_x B_3 + \alpha B_2 = 0, \qquad (9.5)$$
$$\partial_x B_2 - \partial_y B_1 + \alpha B_3 = 0.$$

Solving this system for B_3 leads to the equation

$$\Delta B_3 - \left\langle \frac{\nabla \alpha}{\alpha}, \nabla B_3 \right\rangle + \alpha^2 B_3 = 0 \qquad (9.6)$$

where $\langle \cdot, \cdot \rangle$ denotes the usual scalar product of two vectors.

Note that

$$\alpha \operatorname{div} \left(\frac{1}{\alpha} \nabla B_3 \right) = \Delta B_3 - \left\langle \frac{\nabla \alpha}{\alpha}, \nabla B_3 \right\rangle$$

and hence (9.6) can be rewritten as

$$\operatorname{div} \left(\frac{1}{\alpha} \nabla B_3 \right) + \alpha B_3 = 0. \qquad (9.7)$$

Thus equation (9.1) reduces to an equation of the form (3.10) with $p = 1/\alpha$ and $q = \alpha$.

Let us notice that (see, Proposition 28)

$$\operatorname{div} \frac{1}{\alpha} \nabla + \alpha = \frac{1}{\sqrt{\alpha}} (\Delta - r) \frac{1}{\sqrt{\alpha}}$$

where

$$r = -\frac{1}{2}\frac{\Delta\alpha}{\alpha} + \frac{3}{4}\left(\frac{\nabla\alpha}{\alpha}\right)^2 - \alpha^2. \tag{9.8}$$

That is B_3 is a solution of (9.7) iff the function $f = B_3/\sqrt{\alpha}$ is a solution of the stationary Schrödinger equation

$$(-\Delta + r)\,f = 0 \tag{9.9}$$

with r defined by (9.8). As was explained in Section 3.3, given its particular solution this equation reduces to a Vekua equation of the form (3.15). Unfortunately, in general we are not able to propose a particular solution of (9.7). Nevertheless in an important special case when α depends on one Cartesian variable, a particular solution of (9.7) is always available in explicit form. We give this result in the next section.

9.3 Solution in the case when α is a function of one Cartesian variable

Let us consider equation (9.9) where $\alpha = \alpha(y)$. We assume that α is a nonvanishing function and look for a solution of the corresponding ordinary differential equation

$$\frac{d^2 f}{dy^2} + \left(\frac{1}{2}\frac{\alpha''}{\alpha} - \frac{3}{4}\left(\frac{\alpha'}{\alpha}\right)^2 + \alpha^2\right) f = 0.$$

Its general solution is known (see [56, 2.162 (14)]) and is given by the expression

$$f(y) = \frac{c_1}{\sqrt{\alpha(y)}}\sin\mathcal{A}(y) + \frac{c_2}{\sqrt{\alpha(y)}}\cos\mathcal{A}(y) \tag{9.10}$$

where $\mathcal{A}$ is an antiderivative of α and c_1, c_2 are arbitrary real constants.

Choosing, e.g., $c_1 = 1$, $c_2 = 0$ and calculating the coefficient $(\partial_{\bar{z}} f)/f$ we arrive at the following Vekua equation which is equivalent to (9.1) in the case under consideration (and which is considered in any simply connected domain where $\sin\mathcal{A}(y)$ does not vanish):

$$\partial_{\bar{z}} W(x,y) = \frac{i}{2}\left(\alpha(y)\cot\mathcal{A}(y) - \frac{\alpha'(y)}{2\alpha(y)}\right)\overline{W}(x,y).$$

Note that $F = f = \frac{\sin\mathcal{A}(y)}{\sqrt{\alpha(y)}}$ and $G = \frac{i}{f} = \frac{i\sqrt{\alpha(y)}}{\sin\mathcal{A}(y)}$ represent a generating pair for this Vekua equation and hence if W is its solution, the corresponding pseudoanalytic function of the second kind $\omega = \frac{1}{f}\operatorname{Re}W + if\operatorname{Im}W$ satisfies the equation

$$\omega_{\bar{z}} = \frac{1 - f^2}{1 + f^2}\overline{\omega}_z \tag{9.11}$$

which can be written in the form of the system

$$\phi_x = \frac{1}{f^2}\psi_y, \qquad \phi_y = -\frac{1}{f^2}\psi_x$$

where $\phi = \operatorname{Re}\omega$ and $\psi = \operatorname{Im}\omega$.

For f being representable in a separable form $f(x,y) = X(x)Y(y)$, the formulas for constructing corresponding formal powers explicitly were given in Section 4.2. Using them we obtain the following representation for the formal powers corresponding to (9.11)

$$_*Z^{(n)}(a, z_0; z) = a_1 \sum_{k=0}^{n} \binom{n}{k}(x-x_0)^{(n-k)} i^k Y^k + i a_2 \sum_{k=0}^{n} \binom{n}{k}(x-x_0)^{(n-k)} i^k \widetilde{Y}^k$$

where $z_0 = x_0 + iy_0$ is an arbitrary point of the domain of interest, a is an arbitrary complex number: $a = a_1 + ia_2$, Y^k and $\widetilde{Y}^k$ are constructed as

$$Y^{(0)}(y_0, y) = \widetilde{Y}^{(0)}(y_0, y) = 1$$

and for $n = 1, 2, \ldots$

$$Y^{(n)}(y_0, y) = n \int_{y_0}^{y} Y^{(n-1)}(y_0, \eta) f^2(\eta) d\eta \qquad n \text{ odd,}$$

$$Y^{(n)}(y_0, y) = n \int_{y_0}^{y} Y^{(n-1)}(y_0, \eta) \frac{d\eta}{f^2(\eta)} \qquad n \text{ even,}$$

$$\widetilde{Y}^{(n)}(x_0, x) = n \int_{y_0}^{y} \widetilde{Y}^{(n-1)}(y_0, \eta) \frac{d\eta}{f^2(\eta)} \qquad n \text{ odd,}$$

$$\widetilde{Y}^{(n)}(x_0, x) = n \int_{y_0}^{y} \widetilde{Y}^{(n-1)}(y_0, \eta) f^2(\eta) d\eta \qquad n \text{ even.}$$

The system $\left\{ {}_*Z^{(n)}(1, z_0; z), \, {}_*Z^{(n)}(i, z_0; z) \right\}_{n=0}^{\infty}$ represents a complete (in a compact uniform convergence topology) system of solutions of (9.11), which means that any solution ω of (9.11) in a simply connected domain Ω can be represented as a series

$$\omega(z) = \sum_{n=0}^{\infty} {}_*Z^{(n)}(a_n, z_0; z) = \sum_{n=0}^{\infty} \left(a'_n \, {}_*Z^{(n)}(1, z_0; z) + a''_n \, {}_*Z^{(n)}(i, z_0; z) \right)$$

where $a'_n = \operatorname{Re} a_n$, $a''_n = \operatorname{Im} a_n$ and the series converges normally. Consequently the system of functions

$$\left\{ f(y) \operatorname{Re}({}_*Z^{(n)}(1, z_0; z)), \quad f(y) \operatorname{Re}({}_*Z^{(n)}(i, z_0; z)) \right\}_{n=0}^{\infty}$$

represents in the same sense a complete system of solutions of (9.9) with r defined by (9.8), and

$$\left\{ \sqrt{\alpha(y)} f(y) \operatorname{Re}(_*Z^{(n)}(1, z_0; z)), \quad \sqrt{\alpha(y)} f(y) \operatorname{Re}(_*Z^{(n)}(i, z_0; z)) \right\}_{n=0}^{\infty} \qquad (9.12)$$

is a complete system of solutions of (9.7). Thus in the case under consideration any solution B_3 of (9.7) can be represented in the form

$$B_3(x, y) = \sum_{n=0}^{\infty} (a_n \sin \mathcal{A}(y) \operatorname{Re}(_*Z^{(n)}(1, z_0; z)) + b_n \sin \mathcal{A}(y) \operatorname{Re}(_*Z^{(n)}(i, z_0; z)))$$

where a_n and b_n are real constants.

The other two components of the vector $\vec{B}$ are obtained from (9.4) and (9.5):

$$B_1 = -\frac{1}{\alpha} \partial_y B_3 \quad \text{and} \quad B_2 = \frac{1}{\alpha} \partial_x B_3 \qquad (9.13)$$

that gives us a complete system of solutions of (9.1) in the case under consideration. On the following example we explain how this procedure works.

Example 105. Let us consider the following relatively simple situation in which the corresponding integrals are not difficult to evaluate. Let

$$\alpha(y) = \frac{1}{\sqrt{1 - y^2}} \qquad (9.14)$$

and Ω be an open unitary disk with a center in the origin. We take in (9.10) $c_1 = 0$ and $c_2 = 1$. Then it is easy to verify that

$$f(y) = (1 - y^2)^{\frac{3}{4}}.$$

The first three formal powers with a centre in the origin can be calculated as follows:

$$_*Z^{(1)}(1, 0; z) = x + i \left[\frac{y(1 - y^2)^{\frac{3}{2}}}{4} + \frac{3y(1 - y^2)^{\frac{1}{2}}}{8} + \frac{3}{8} \arcsin y \right],$$

$$_*Z^{(1)}(i, 0; z) = -\frac{y}{(1 - y^2)^{\frac{1}{2}}} + ix,$$

$$_*Z^{(2)}(1, 0; z) = x^2 - \frac{1}{4} y^2 - \frac{3}{4} \frac{y \arcsin y}{(1 - y^2)^{\frac{1}{2}}}$$

$$+ 2ix \left(\frac{y(1 - y^2)^{\frac{3}{2}}}{4} + \frac{3y(1 - y^2)^{\frac{1}{2}}}{8} + \frac{3}{8} \arcsin y \right),$$

$$_*Z^{(2)}(i, 0; z) = -\frac{2xy}{(1 - y^2)^{\frac{1}{2}}} + i \left(x^2 - y^2 - \frac{1}{2} y^4 \right),$$

$$_*Z^{(3)}(1,0;z) = x^3 - 3x\left(\frac{1}{4}y^2 + \frac{3}{4}\frac{y\arcsin y}{(1-y^2)^{\frac{1}{2}}}\right)$$

$$+ 3ix^2\left(\frac{y(1-y^2)^{\frac{3}{2}}}{4} + \frac{3y(1-y^2)^{\frac{1}{2}}}{8} + \frac{3}{8}\arcsin y\right)$$

$$- i\left(-\frac{3}{24}y(1-y^2)^{\frac{5}{2}} + \frac{3}{96}y(1-y^2)^{\frac{3}{2}} + y(1-y^2)^{\frac{1}{2}}\left(\frac{51}{128} - \frac{9}{64}y^2\right)\right.$$

$$\left. - \frac{9}{16}(1-y^2)^2\arcsin y + \frac{33}{128}\arcsin y\right),$$

$$_*Z^{(3)}(i,0;z) = -\frac{3x^2y}{(1-y^2)^{\frac{1}{2}}} + \frac{3}{4}\frac{y(1+y^2)}{(1-y^2)^{\frac{1}{2}}} - \frac{3}{4}\arcsin y + ix\left(x^2 - 3\left(y^2 - \frac{1}{2}y^4\right)\right).$$

Now taking the real parts of these formal powers and multiplying them by the factor $\sqrt{\alpha}f$ (see (9.12)) we obtain the first elements of the complete system of solutions of (9.7), that is any solution B_3 of (9.7) in a simply connected domain can be represented as an infinite linear combination of the functions

$$\left\{(1-y^2)^{\frac{1}{2}}, \quad x(1-y^2)^{\frac{1}{2}}, \quad -y, \quad (1-y^2)^{\frac{1}{2}}\left(x^2 - \frac{1}{4}y^2 - \frac{3}{4}\frac{y\arcsin y}{(1-y^2)^{\frac{1}{2}}}\right),\right.$$

$$- 2xy, \quad (1-y^2)^{\frac{1}{2}}\left(x^3 - 3x\left(\frac{1}{4}y^2 + \frac{3}{4}\frac{y\arcsin y}{(1-y^2)^{\frac{1}{2}}}\right)\right),$$

$$\left. - 3x^2y + \frac{3}{4}y(1+y^2) - \frac{3}{4}(1-y^2)^{\frac{1}{2}}\arcsin y, \dots\right\}$$

and the corresponding series converges normally.

From (9.13) it is easy to calculate the corresponding components B_1 and B_2 respectively,

$$\left\{y, \quad xy, \quad (1-y^2)^{\frac{1}{2}}, \quad \frac{3}{4}(1-y^2)^{\frac{1}{2}}\arcsin y + y\left(x^2 - \frac{3}{4}y^2 + \frac{5}{4}\right),\right.$$

$$2x(1-y^2)^{\frac{1}{2}}, \quad \frac{9}{4}x(1-y^2)^{\frac{1}{2}}\arcsin y + y\left(x^3 - \frac{9}{4}xy^2 + \frac{15}{4}x\right),$$

$$\left. - (\frac{9}{4}y^2 - 3x^2)(1-y^2)^{\frac{1}{2}} - \frac{3}{4}y\arcsin y, \dots\right\}$$

and

$$\left\{0, \quad (1-y^2), \quad 0, \quad 2x(1-y^2), \quad -2y(1-y^2)^{\frac{1}{2}},\right.$$

$$\left. (3x^2 - \frac{3}{4}y^2)(1-y^2) - \frac{9}{4}y(1-y^2)^{\frac{1}{2}}\arcsin y, \quad -6xy(1-y^2)^{\frac{1}{2}}, \dots\right\}.$$

Thus, we obtain the following complete system of solutions of (9.1) with the proportionality factor α defined by (9.14):

$$\vec{B}_0 = \begin{pmatrix} y \\ 0 \\ (1-y^2)^{\frac{1}{2}} \end{pmatrix}, \quad \vec{B}_1 = \begin{pmatrix} xy \\ (1-y^2) \\ x(1-y^2)^{\frac{1}{2}} \end{pmatrix}, \quad \vec{B}_2 = \begin{pmatrix} (1-y^2)^{\frac{1}{2}} \\ 0 \\ -y \end{pmatrix},$$

$$\vec{B}_3 = \begin{pmatrix} \frac{3}{4}(1-y^2)^{\frac{1}{2}} \arcsin y + y(x^2 - \frac{3}{4}y^2 + \frac{5}{4}) \\ 2x(1-y^2) \\ (1-y^2)^{\frac{1}{2}}\left(x^2 - \frac{1}{4}y^2 - \frac{3}{4}\frac{y \arcsin y}{(1-y^2)^{\frac{1}{2}}} \right) \end{pmatrix}, \quad \vec{B}_4 = \begin{pmatrix} 2x(1-y^2)^{\frac{1}{2}} \\ -2y(1-y^2)^{\frac{1}{2}} \\ -2xy \end{pmatrix},$$

$$\vec{B}_5 = \begin{pmatrix} \frac{9}{4}x(1-y^2)^{\frac{1}{2}} \arcsin y + y\left(x^3 - \frac{9}{4}xy^2 + \frac{15}{4}x \right) \\ (3x^2 - \frac{3}{4}y^2)(1-y^2) - \frac{9}{4}y(1-y^2)^{\frac{1}{2}} \arcsin y \\ (1-y^2)^{\frac{1}{2}}\left(x^3 - 3x\left(\frac{1}{4}y^2 + \frac{3}{4}\frac{y \arcsin y}{(1-y^2)^{\frac{1}{2}}} \right) \right) \end{pmatrix},$$

$$\vec{B}_6 = \begin{pmatrix} -(\frac{9}{4}y^2 - 3x^2)(1-y^2)^{\frac{1}{2}} - \frac{3}{4}y \arcsin y \\ -6xy(1-y^2)^{\frac{1}{2}} \\ -3x^2 y + \frac{3}{4}y(1+y^2) - \frac{3}{4}(1-y^2)^{\frac{1}{2}} \arcsin y \end{pmatrix},$$

$$\cdots$$

Of course the integrals involved in the construction of the complete system of solutions are not always sufficiently easy to evaluate explicitly as in Example 105. Nevertheless our numerical experiments confirm that in general the formal powers and hence the solutions of (9.1) can be calculated with a remarkable accuracy. For example, the vector $\vec{B}_{40}$ (see notations in Example 105) in the Matlab 7 package on a usual PC can be calculated with a precision of order 10^{-4}. Thus, the use of formal powers for numerical solution of boundary value problems corresponding to (9.1) and more generally to equations of the form (3.10) is really promising.

Chapter 10

Static Maxwell System in Axially Symmetric Inhomogeneous Media

10.1 Meridional and transverse fields

Consider the static Maxwell system

$$\operatorname{div}(\varepsilon \mathbf{E}) = 0, \qquad \operatorname{rot} \mathbf{E} = 0 \tag{10.1}$$

where we suppose that ε is a function of the cylindrical radial variable $r = \sqrt{x_1^2 + x_2^2}$: $\varepsilon = \varepsilon(r)$. Two important situations are usually studied: the meridional field and the transverse field.

The first case is characterized by the condition that the vector $\mathbf{E}$ is independent of the angular coordinate θ and the component E_θ of the vector $\mathbf{E}$ in cylindrical coordinates vanishes identically. A vector of such a field belongs to a plane containing the axis x_3 and depends only on the distance r to this axis as well as on the coordinate x_3. The field then is completely described by a two-component vector-function in the plane (r, x_3).

The second case is characterized by the condition that the vector $\mathbf{E}$ is independent of x_3 and the component E_3 is identically zero. A vector of such a field belongs to a plane perpendicular to the axis x_3 and the corresponding model reduces to a two-component vector-function in the plane (x_1, x_2).

In both cases following [61] we construct a complete system of solutions of the corresponding model. We use the fact that in both cases the system (10.1) reduces to a system describing p-analytic functions

$$u_x = \frac{1}{p} v_y, \qquad u_y = -\frac{1}{p} v_x. \tag{10.2}$$

In the first case the function p is a function of one Cartesian variable x, meanwhile in the second it is a function of $r = \sqrt{x^2 + y^2}$. In both cases we construct an infinite system of formal powers.

10.2   Reduction of the static Maxwell system to p-analytic functions

10.2.1   The meridional case

Introducing the cylindrical coordinates and making the assumptions that $\mathbf{E}$ is independent of the angular variable θ and that the component E_θ is identically zero, we obtain that (10.1) can be written as

$$\frac{\partial E_r}{\partial x_3} - \frac{\partial E_3}{\partial r} = 0, \qquad \frac{1}{r}\frac{\partial (r\varepsilon E_r)}{\partial r} + \frac{\partial (\varepsilon E_3)}{\partial x_3} = 0.$$

Set $x = r$, $y = x_3$, $u = E_3$ and $v = r\varepsilon E_r$. Then the system takes the form

$$u_x = \frac{1}{x\varepsilon(x)}v_y, \qquad u_y = -\frac{1}{x\varepsilon(x)}v_x,$$

where the subindices denote the derivatives with respect to the corresponding variables. Thus, in the case of a meridional field the vector $\mathbf{E}$ is completely described by an $x\varepsilon(x)$-analytic function $\omega = u + iv$.

10.2.2   The transverse case

We assume that $\mathbf{E}$ is independent of the longitudinal variable x_3 and $E_3 \equiv 0$. Then from (10.1) we have that the vector $(E_1, E_2)^T$ is the gradient of a function $u = u(x_1, x_2)$ which satisfies the two-dimensional conductivity equation

$$\operatorname{div}(\varepsilon \nabla u) = 0. \tag{10.3}$$

Write $x = x_1$, $y = x_2$, $z = x + iy$ and consider the system

$$u_x = \frac{1}{\varepsilon}v_y, \qquad u_y = -\frac{1}{\varepsilon}v_x. \tag{10.4}$$

It is easy to see that if the function $\omega = u + iv$ is its solution, then u is a solution of (10.3), and vice versa, if u is a solution of (10.3) in a simply connected domain Ω then choosing

$$v = \overline{A}(i\varepsilon u_{\overline{z}}), \tag{10.5}$$

we obtain that $\omega = u + iv$ is a solution of (10.4). Note that v is a solution of the equation

$$\operatorname{div}(\frac{1}{\varepsilon}\nabla v) = 0.$$

Thus, equation (10.3) (and hence the system (10.1) in the case under consideration) is equivalent to the system (10.4) in the sense that if $\omega = u + iv$ is a solution of (10.4) then its real part u is a solution of (10.3) and vice versa, if u is a solution of (10.3) then $\omega = u + iv$, where v is constructed according to (10.5) is a solution of (10.4).

We reduced both considered cases, the meridional and the transverse, to the system describing p-analytic functions. In the first case $p = x\varepsilon(x)$ is a function of one Cartesian variable and in the second $p = \varepsilon(r)$, $r = \sqrt{x^2 + y^2}$. As we show below in both cases we are able to construct explicitly a complete system of formal powers and hence a complete system of exact solutions of the corresponding Maxwell system. Let us notice that equation (10.3) with ε being a function of the variable r was considered in the recent work [35] with applications to electrical impedance tomography. The algorithm proposed in that work implies numerical solution of a number of ordinary differential equations arising after a standard separation of variables. Our construction of a complete system of solutions of (10.3) is based on essentially different ideas and does not require solving numerically any differential equation.

10.3 Construction of formal powers

10.3.1 Formal powers in the meridional case

As was shown above in the meridional case the Maxwell system reduces to the pair of equations

$$u_x = \frac{1}{x\varepsilon(x)} v_y, \qquad u_y = -\frac{1}{x\varepsilon(x)} v_x$$

which is equivalent to the system

$$\sigma(x)\phi_x = \tau(y)\psi_y, \qquad \sigma(x)\phi_y = -\tau(y)\psi_x.$$

Taking $\sigma(x) = x\varepsilon(x)$ and $\tau \equiv 1$ we can use L. Bers' elegant formulas from Section 4.2. Let

$$X^{(0)}(x_0, x) = \widetilde{X}^{(0)}(x_0, x) = 1$$

and for $n = 1, 2, \ldots$ let

$$X^{(n)}(x_0, x) = n \int_{x_0}^{x} X^{(n-1)}(x_0, t) \frac{1}{t\varepsilon(t)} dt \qquad \text{for odd } n,$$

$$X^{(n)}(x_0, x) = n \int_{x_0}^{x} X^{(n-1)}(x_0, t) t\varepsilon(t) dt \qquad \text{for even } n,$$

$$\widetilde{X}^{(n)}(x_0, x) = n \int_{x_0}^{x} \widetilde{X}^{(n-1)}(x_0, t) t\varepsilon(t) dt \qquad \text{for odd } n,$$

$$\widetilde{X}^{(n)}(x_0, x) = n \int\limits_{x_0}^{x} \widetilde{X}^{(n-1)}(x_0, t) \frac{1}{t\varepsilon(t)} dt \qquad \text{for even } n.$$

Then the formal powers in the meridional case are given by the expressions

$$_*Z^{(n)}(a' + ia'', z_0; z) = a' \sum_{k=0}^{n} \binom{n}{k} X^{(n-k)} i^k (y - y_0)^k$$

$$+ ia'' \sum_{k=0}^{n} \binom{n}{k} \widetilde{X}^{(n-k)} i^k (y - y_0)^k \qquad \text{for odd } n$$

and

$$_*Z^{(n)}(a' + ia'', z_0; z) = a' \sum_{k=0}^{n} \binom{n}{k} \widetilde{X}^{(n-k)} i^k (y - y_0)^k$$

$$+ ia'' \sum_{k=0}^{n} \binom{n}{k} X^{(n-k)} i^k (y - y_0)^k \qquad \text{for even } n.$$

10.3.2   Formal powers in the transverse case

This case reduces to the system

$$u_x = \frac{1}{\varepsilon} v_y, \qquad u_y = -\frac{1}{\varepsilon} v_x$$

where ε is a positive differentiable function of $r = \sqrt{x^2 + y^2}$. The system describing ε-analytic functions is equivalent to the main Vekua equation (3.15) where $f = \sqrt{\varepsilon}$. In order to apply Theorem 77 we let $u = \ln r$ and $U(u) = \sqrt{\varepsilon(e^u)}$. Then taking $V \equiv 1$ we obtain the generating pair (F, G) for equation (3.15) in the desirable form

$$F = U(u), \qquad G = \frac{i}{U(u)}. \tag{10.6}$$

The analytic function Φ (from Theorem 77) corresponding to the polar coordinate system has the form $\Phi(z) = \ln z$ and consequently $\Phi_z(z) = 1/z$. We note that Φ_z has a pole in the origin and a zero at infinity. Thus, Theorem 77 is applicable in any domain Ω which does not include these two points. Moreover, as for constructing formal powers we need to use the recursive integration defined by (4.1); in what follows we require Ω to be any bounded simply connected domain not containing the origin.

From Theorem 77 we have that a generating sequence corresponding to the generating pair (10.6) can be defined as

$$F_m = \frac{U}{z^m} \quad \text{and} \quad G_m = \frac{i}{z^m U} \qquad \text{for even } m$$

and

$$F_m = \frac{1}{z^m U} \quad \text{and} \quad G_m = \frac{iU}{z^m} \quad \text{for odd } m.$$

In order to have a complete system of formal powers for each n we need to construct $Z^{(n)}(1, z_0; z)$ and $Z^{(n)}(i, z_0; z)$.

For $n = 0$ we have

$$Z^{(0)}(1, z_0; z) = \lambda_1^{(0)} F(z) + \mu_1^{(0)} G(z)$$

and

$$Z^{(0)}(i, z_0; z) = \lambda_i^{(0)} F(z) + \mu_i^{(0)} G(z)$$

where $\lambda_1^{(0)}$, $\mu_1^{(0)}$ are real constants chosen so that

$$\lambda_1^{(0)} F(z_0) + \mu_1^{(0)} G(z_0) = 1$$

and $\lambda_i^{(0)}$, $\mu_i^{(0)}$ are real constants such that

$$\lambda_i^{(0)} F(z_0) + \mu_i^{(0)} G(z_0) = i.$$

Taking into account that F is real and G is imaginary we obtain that

$$\lambda_1^{(0)} = \frac{1}{F(z_0)}, \qquad \mu_1^{(0)} = 0,$$

$$\lambda_i^{(0)} = 0, \qquad \mu_i^{(0)} = F(z_0).$$

Thus,

$$Z^{(0)}(1, z_0; z) = \frac{F(z)}{F(z_0)} = \sqrt{\frac{\varepsilon(r)}{\varepsilon(r_0)}}$$

and

$$Z^{(0)}(i, z_0; z) = \frac{iF(z_0)}{F(z)} = i\sqrt{\frac{\varepsilon(r_0)}{\varepsilon(r)}}$$

where $r_0 = |z_0|$.

For constructing $Z^{(1)}(1, z_0; z)$ and $Z^{(1)}(i, z_0; z)$ we need first the formal powers $Z_1^{(0)}(1, z_0; z)$ and $Z_1^{(0)}(i, z_0; z)$. According to Definition 52 they have the form

$$Z_1^{(0)}(1, z_0; z) = \lambda_1^{(1)} F_1(z) + \mu_1^{(1)} G_1(z)$$

and

$$Z_1^{(0)}(i, z_0; z) = \lambda_i^{(1)} F_1(z) + \mu_i^{(1)} G_1(z)$$

where $\lambda_1^{(1)}$, $\mu_1^{(1)}$ are real numbers such that

$$\lambda_1^{(1)} F_1(z_0) + \mu_1^{(1)} G_1(z_0) = 1$$

and $\lambda_i^{(1)}$, $\mu_i^{(1)}$ are real numbers such that

$$\lambda_i^{(1)} F_1(z_0) + \mu_i^{(1)} G_1(z_0) = i.$$

Thus in order to determine $\lambda_1^{(1)}$, $\mu_1^{(1)}$ and $\lambda_i^{(1)}$, $\mu_i^{(1)}$ we should solve two systems of linear algebraic equations:

$$\lambda_1^{(1)} \frac{1}{z_0 \varepsilon^{1/2}(r_0)} + \mu_1^{(1)} \frac{i\varepsilon^{1/2}(r_0)}{z_0} = 1$$

and

$$\lambda_i^{(1)} \frac{1}{z_0 \varepsilon^{1/2}(r_0)} + \mu_i^{(1)} \frac{i\varepsilon^{1/2}(r_0)}{z_0} = i$$

which can be rewritten as

$$\lambda_1^{(1)} + \mu_1^{(1)} i\varepsilon(r_0) = \varepsilon^{1/2}(r_0) z_0$$

and

$$\lambda_i^{(1)} + \mu_i^{(1)} i\varepsilon(r_0) = i\varepsilon^{1/2}(r_0) z_0.$$

From here we obtain

$$\lambda_1^{(1)} = \varepsilon^{1/2}(r_0) x_0, \qquad \mu_1^{(1)} = \varepsilon^{-1/2}(r_0) y_0,$$
$$\lambda_i^{(1)} = -\varepsilon^{1/2}(r_0) y_0, \quad \mu_i^{(1)} = \varepsilon^{-1/2}(r_0) x_0.$$

Let us notice that in general for odd m we have

$$Z_m^{(0)}(1, z_0; z) = \frac{\lambda_1^{(m)}}{z^m \varepsilon^{1/2}(r)} + \frac{i\mu_1^{(m)} \varepsilon^{1/2}(r)}{z^m},$$

$$Z_m^{(0)}(i, z_0; z) = \frac{\lambda_i^{(m)}}{z^m \varepsilon^{1/2}(r)} + \frac{i\mu_i^{(m)} \varepsilon^{1/2}(r)}{z^m},$$

where

$$\lambda_1^{(m)} = \varepsilon^{1/2}(r_0) \operatorname{Re} z_0^m = \varepsilon^{1/2}(r_0) r_0^m \cos m\theta_0,$$

$$\mu_1^{(m)} = \varepsilon^{-1/2}(r_0) \operatorname{Im} z_0^m = \varepsilon^{-1/2}(r_0) r_0^m \sin m\theta_0,$$

$$\lambda_i^{(m)} = -\varepsilon^{1/2}(r_0) \operatorname{Im} z_0^m = -\varepsilon^{1/2}(r_0) r_0^m \sin m\theta_0,$$

$$\mu_i^{(m)} = \varepsilon^{-1/2}(r_0) \operatorname{Re} z_0^m = \varepsilon^{-1/2}(r_0) r_0^m \cos m\theta_0,$$

θ_0 is the argument of the complex number z_0.

Thus, for odd m:

$$Z_m^{(0)}(1, z_0; z) = \left(\frac{r_0}{z}\right)^m \left(\cos m\theta_0 \sqrt{\frac{\varepsilon(r_0)}{\varepsilon(r)}} + i\sin m\theta_0 \sqrt{\frac{\varepsilon(r)}{\varepsilon(r_0)}}\right),$$

$$Z_m^{(0)}(i, z_0; z) = \left(\frac{r_0}{z}\right)^m \left(-\sin m\theta_0 \sqrt{\frac{\varepsilon(r_0)}{\varepsilon(r)}} + i\cos m\theta_0 \sqrt{\frac{\varepsilon(r)}{\varepsilon(r_0)}}\right).$$

In a similar way we obtain the corresponding formulas for even m:

$$Z_m^{(0)}(1, z_0; z) = \left(\frac{r_0}{z}\right)^m \left(\cos m\theta_0 \sqrt{\frac{\varepsilon(r)}{\varepsilon(r_0)}} + i\sin m\theta_0 \sqrt{\frac{\varepsilon(r_0)}{\varepsilon(r)}}\right),$$

$$Z_m^{(0)}(i, z_0; z) = \left(\frac{r_0}{z}\right)^m \left(-\sin m\theta_0 \sqrt{\frac{\varepsilon(r)}{\varepsilon(r_0)}} + i\cos m\theta_0 \sqrt{\frac{\varepsilon(r_0)}{\varepsilon(r)}}\right).$$

In order to apply formula (4.1) for constructing formal powers of higher orders we need to calculate the adjoint generating pairs (F_m^*, G_m^*). For odd m we have

$$F_m^* = -\frac{iz^m}{\varepsilon^{1/2}(r)}, \quad G_m^* = \varepsilon^{1/2}(r)z^m.$$

For even m we obtain

$$F_m^* = -iz^m\varepsilon^{1/2}(r), \quad G_m^* = \frac{z^m}{\varepsilon^{1/2}(r)}.$$

Now the whole procedure of construction of formal powers can be easily algorithmized according to the formula (4.1). The obtained system of formal powers

$$\left\{Z^{(n)}(1, z_0; z), \quad Z^{(n)}(i, z_0; z)\right\}_{n=0}^{\infty}$$

is complete in the space of all solutions of the main Vekua equation (3.15) with $f = \varepsilon^{1/2}(r)$, i.e., any regular solution W of (3.15) in Ω can be represented in the form of a normally convergent series

$$W(z) = \sum_{n=0}^{\infty} Z^{(n)}(a_n, z_0; z) = \sum_{n=0}^{\infty} \left(a_n' Z^{(n)}(1, z_0; z) + a_n'' Z^{(n)}(i, z_0; z)\right)$$

where $a_n' = \operatorname{Re} a_n$, $a_n'' = \operatorname{Im} a_n$ and z_0 is an arbitrary fixed point in Ω.

Part IV

Hyperbolic Pseudoanalytic Functions

Chapter 11

Hyperbolic Numbers and Analytic Functions

In this part of the book we use the results and follow the exposition from [80] where a hyperbolic analogue of pseudoanalytic function theory was developed which proves to be extremely useful for studying hyperbolic partial differential equations. We show that solutions of the Klein-Gordon equation with an arbitrary potential are closely related to certain hyperbolic pseudoanalytic functions, the result of a factorization of the Klein-Gordon operator with the aid of two Vekua-type operators. As one of the corollaries we obtain a method for explicit construction of infinite systems of solutions of the considered Klein-Gordon equation. Our approach is based on the application of the algebra of hyperbolic numbers [110, 114] instead of that of complex numbers and generalizes some earlier works dedicated to hyperbolic analytic function theory [84, 95, 52]. It should be mentioned that the elliptic and hyperbolic pseudoanalytic function theories naturally are quite different. Nevertheless as we show following [80] there are many important common features.

It has been proven (see, e.g., [57]) that there exist essentially three possible ways to generalize real numbers into real algebras of dimension 2. Indeed, each possible system can be reduced to one of the following:

1. numbers $a + b\mathrm{i}$ with $\mathrm{i}^2 = -1$ (complex numbers);
2. numbers $a + b\mathrm{j}$ with $\mathrm{j}^2 = 1$ (hyperbolic numbers);
3. numbers $a + b\mathrm{k}$ with $\mathrm{k}^2 = 0$ (dual numbers).

Here we use the set of hyperbolic numbers, also called duplex numbers (see, e.g., [110, 114]) and denote it by

$$\mathbb{D} := \left\{ x + t\mathrm{j} \ : \ \mathrm{j}^2 = 1, \ x, t \in \mathbb{R} \right\} \cong \mathrm{Cl}_{\mathbb{R}}(0, 1). \tag{11.1}$$

It is easy to see that this algebra of hyperbolic numbers is commutative and contains zero divisors.

As in the case of complex numbers, we denote the real and "imaginary" parts of $z = x + t\mathrm{j} \in \mathbb{D}$ by $x = \mathrm{Re}(z)$ and $t = \mathrm{Im}(z)$. Now defining the conjugate as $\bar{z} := x - t\mathrm{j}$ and the hyperbolic modulus as $|z|^2 := z\bar{z} = x^2 - t^2$, we can verify that the inverse of z whenever it exists is given by

$$z^{-1} = \frac{\bar{z}}{|z|^2}. \tag{11.2}$$

From this, we find that the set $\mathcal{NC}$ of zero divisors of $\mathbb{D}$, called the *null-cone*, is given by

$$\mathcal{NC} = \big\{ x + t\mathrm{j} \ : \ |x| = |t| \big\}.$$

It is also possible to define differentiability of a function at a point of $\mathbb{D}$ [105, 108]:

Definition 106. Let U be an open set of $\mathbb{D}$ and $z_0 \in U$. Then, $f : U \subseteq \mathbb{D} \longrightarrow \mathbb{D}$ is said to be $\mathbb{D}$-differentiable at z_0 with derivative equal to $f'(z_0) \in \mathbb{D}$ if

$$\lim_{\substack{z \to z_0 \\ (z-z_0 \text{ inv.})}} \frac{f(z) - f(z_0)}{z - z_0} = f'(z_0). \tag{11.3}$$

Here z tends to z_0 following the invertible trajectories. We also say that the function f is $\mathbb{D}$-holomorphic on an open set U if and only if f is $\mathbb{D}$-differentiable at each point of U.

Any hyperbolic number can be seen as an element of $\mathbb{R}^2$, so a function $f(x + t\mathrm{j}) = f_1(x,t) + f_2(x,t)\mathrm{j}$ can be seen as a mapping $f(x,t) = \big(f_1(x,t), f_2(x,t)\big)$ of $\mathbb{R}^2$.

Theorem 107. *Let U be an open set and $f : U \subseteq \mathbb{D} \longrightarrow \mathbb{D}$ such that $f \in C^1(U)$. Let also $f(x + t\mathrm{j}) = f_1(x,t) + f_2(x,t)\mathrm{j}$. Then f is $\mathbb{D}$-holomorphic on U if and only if*

$$\frac{\partial f_1}{\partial x} = \frac{\partial f_2}{\partial t} \qquad and \qquad \frac{\partial f_2}{\partial x} = \frac{\partial f_1}{\partial t}. \tag{11.4}$$

Moreover $f' = \dfrac{\partial f_1}{\partial x} + \dfrac{\partial f_2}{\partial x}\mathrm{j}$ and $f'(z)$ is invertible if and only if $\det \mathcal{J}_f(z) \neq 0$, where $\mathcal{J}_f(z)$ is the Jacobian matrix of f at z.

Every hyperbolic number $x + t\mathrm{j}$ has the unique idempotent representation

$$x + t\mathrm{j} = (x+t)\mathrm{e}_1 + (x-t)\mathrm{e}_2, \tag{11.5}$$

where $\mathrm{e}_1 = \dfrac{1+\mathrm{j}}{2}$ and $\mathrm{e}_2 = \dfrac{1-\mathrm{j}}{2}$. This representation is useful because with its aid addition, multiplication and division can be done term-by-term.

The notion of holomorphicity can also be seen using this notation. We define the projections $P_1, P_2 : \mathbb{D} \longrightarrow \mathbb{R}$ as $P_1(z) = x + t$ and $P_2(z) = x - t$, where $z = x + t\mathrm{j}$ as well as the following definition.

Definition 108. We say that $X \subseteq \mathbb{D}$ is a $\mathbb{D}$-cartesian set determined by X_1 and X_2 if

$$X = X_1 \times_e X_2 := \{x + tj \in \mathbb{D} : x + tj = w_1 e_1 + w_2 e_2, (w_1, w_2) \in X_1 \times X_2\}. \quad (11.6)$$

It is easy to show that if X_1 and X_2 are open domains of $\mathbb{R}$, then $X_1 \times_e X_2$ is also an open domain of $\mathbb{D}$. Now, it is possible to formulate the following theorem.

Theorem 109. *If* $f_{e_1} : X_1 \longrightarrow \mathbb{R}$ *and* $f_{e_2} : X_2 \longrightarrow \mathbb{R}$ *are real differentiable functions on the open domains* X_1 *and* X_2 *respectively, then the function* $f : X_1 \times_e X_2 \longrightarrow \mathbb{D}$ *defined as*

$$f(x + tj) = f_{e_1}(x + t)e_1 + f_{e_2}(x - t)e_2, \ \forall \ x + tj \in X_1 \times_e X_2 \quad (11.7)$$

is $\mathbb{D}$-*holomorphic on the domain* $X_1 \times_e X_2$ *and*

$$f'(x + tj) = f'_{e_1}(x + t)e_1 + f'_{e_2}(x - t)e_2, \ \forall \ x + tj \in X_1 \times_e X_2. \quad (11.8)$$

Chapter 12

Hyperbolic Pseudoanalytic Functions

12.1 Differential operators

We will consider the variable $z = x + tj$, where x and t are real variables and the corresponding formal differential operators

$$\partial_z = \frac{1}{2}\left(\partial_x + j\partial_t\right) \text{ and } \partial_{\bar{z}} = \frac{1}{2}\left(\partial_x - j\partial_t\right). \tag{12.1}$$

Notation $f_{\bar{z}}$ or f_z means the application of $\partial_{\bar{z}}$ or ∂_z respectively to a hyperbolic function $f(z) = u(z) + v(z)j$. These hyperbolic operators act on sums, products, etc. just as an ordinary derivative and we have the following result in the hyperbolic function theory. We note that

$$f_z = \frac{1}{2}\Big((u_x + v_t) + (v_x + u_t)j\Big) \quad \text{and} \quad f_{\bar{z}} = \frac{1}{2}\Big((u_x - v_t) + (v_x - u_t)j\Big).$$

In view of these operators,

$$f_z(z) = 0 \quad \Leftrightarrow \quad (u_x + v_t) + (v_x + u_t)j = 0 \tag{12.2}$$

i.e., $u_x = -v_t$, $v_x = -u_t$ and

$$f_{\bar{z}}(z) = 0 \quad \Leftrightarrow \quad (u_x + v_t) + (v_x + u_t)j = 0 \tag{12.3}$$

i.e., $u_x = v_t$, $v_x = u_t$.

Lemma 110. *Let $f(x + tj) = u(x, t) + v(x, t)j$ be a hyperbolic function where u_x, u_t, v_x and v_t exist, and are continuous in a neighborhood of z_0. The derivative*

$$f'(z_0) = \lim_{\substack{z \to z_0 \\ (z - z_0 \text{ inv.})}} \frac{f(z) - f(z_0)}{z - z_0} \tag{12.4}$$

exists, if and only if

$$f_{\bar{z}}(z_0) = 0. \tag{12.5}$$

Moreover, $f'(z_0) = f_z(z_0)$ and $f'(z_0)$ is invertible if and only if $\det \mathcal{J}_f(z_0) \neq 0$.

12.2   Hyperbolic pseudoanalytic function theory

Let $z = x + t\mathrm{j}$ where $x, t \in \mathbb{R}$. The theory is based on assigning the part played by 1 and j to two essentially arbitrary hyperbolic functions F and G. We assume that these functions are defined and twice-continuously differentiable in some open domain $\Omega \subset \mathbb{D}$. We require that

$$\mathrm{Im}\{\overline{F(z)}G(z)\} \neq 0. \tag{12.6}$$

Under this condition, (F, G) will be called a generating pair in Ω. Notice that $\mathrm{Im}\{\overline{F(z)}G(z)\} = \begin{vmatrix} \mathrm{Re}\{F(z)\} & \mathrm{Re}\{G(z)\} \\ \mathrm{Im}\{F(z)\} & \mathrm{Im}\{G(z)\} \end{vmatrix}$. It follows, from Cramer's theorem, that for every z_0 in Ω we can find unique constants $\lambda_0, \mu_0 \in \mathbb{R}$ such that $w(z_0) = \lambda_0 F(z_0) + \mu_0 G(z_0)$. More generally we have the following result.

Theorem 111. *Let (F, G) be a generating pair in some open domain Ω. If $w(z) : \Omega \subset \mathbb{D} \to \mathbb{D}$, then there exist* unique *functions $\phi(z), \psi(z) : \Omega \subset \mathbb{D} \to \mathbb{R}$ such that*

$$w(z) = \phi(z)F(z) + \psi(z)G(z), \ \forall z \in \Omega. \tag{12.7}$$

Moreover, we have the following explicit formulas for ϕ and ψ:

$$\phi(z) = \frac{\mathrm{Im}[\overline{w(z)}G(z)]}{\mathrm{Im}[\overline{F(z)}G(z)]}, \ \psi(z) = -\frac{\mathrm{Im}[\overline{w(z)}F(z)]}{\mathrm{Im}[\overline{F(z)}G(z)]}. \tag{12.8}$$

Proof. Let (F, G) be a generating pair in some open domain Ω. Let $z_0 \in \Omega$ with $w(z_0) = x_1 + t_1\mathrm{j}$, $F(z_0) = x_2 + t_2\mathrm{j}$ and $G(z_0) = x_3 + t_3\mathrm{j}$. In this case, $w(z_0) = \phi(z_0)F(z_0) + \psi(z_0)G(z_0)$ with $\phi(z_0), \psi(z_0) \in \mathbb{R}$ if and only if $x_1 = \phi(z_0)x_2 + \psi(z_0)x_3$ and $t_1 = \phi(z_0)t_2 + \psi(z_0)t_3$. That is we obtain the system $AX = B$ where $A = \left(\begin{smallmatrix} x_2 & x_3 \\ t_2 & t_3 \end{smallmatrix} \right)$, $B = \left(\begin{smallmatrix} x_1 \\ t_1 \end{smallmatrix} \right)$ and $X = \left(\begin{smallmatrix} \phi(z_0) \\ \psi(z_0) \end{smallmatrix} \right)$ and the unique solution is $X = A^{-1}B$ where $A^{-1} = \frac{1}{\det A} \left(\begin{smallmatrix} t_3 & -x_3 \\ -t_2 & x_2 \end{smallmatrix} \right)$. Hence,

$$
\begin{aligned}
X &= \frac{1}{\mathrm{Im}[\overline{F(z_0)}G(z_0)]} \begin{pmatrix} t_3 & -x_3 \\ -t_2 & x_2 \end{pmatrix} \begin{pmatrix} x_1 \\ t_1 \end{pmatrix} \\
&= \frac{1}{\mathrm{Im}[\overline{F(z_0)}G(z_0)]} \begin{pmatrix} \mathrm{Im}[\overline{w(z_0)}G(z_0)] \\ -\mathrm{Im}[\overline{w(z_0)}F(z_0)] \end{pmatrix}.
\end{aligned}
\tag{12.9}
$$

Then

$$\phi(z) = \frac{\mathrm{Im}[\overline{w(z)}G(z)]}{\mathrm{Im}[\overline{F(z)}G(z)]}, \ \psi(z) = -\frac{\mathrm{Im}[\overline{w(z)}F(z)]}{\mathrm{Im}[\overline{F(z)}G(z)]}, \ \forall z \in \Omega. \tag{12.10}$$

$\square$

Consequently, every hyperbolic function w defined in some subdomain of Ω admits the unique representation $w = \phi F + \psi G$ where the functions ϕ and ψ are real-valued. Thus, the pair (F, G) generalizes the pair $(1, \mathrm{j})$ which corresponds to hyperbolic analytic function theory. Sometimes it is convenient to associate with the function w the function $\omega = \phi + \mathrm{j}\psi$. The correspondence between w and ω is one-to-one.

We say that $w : \Omega \subset \mathbb{D} \to \mathbb{D}$ possesses at z_0 the (F, G)-derivative $\dot{w}(z_0)$ if the (finite) limit

$$\dot{w}(z_0) = \lim_{\substack{z \to z_0 \\ (z - z_0 \text{ inv.})}} \frac{w(z) - \lambda_0 F(z) - \mu_0 G(z)}{z - z_0} \tag{12.11}$$

exists.

The following expressions are called the characteristic coefficients of the pair (F, G):

$$a_{(F,G)} = -\frac{\bar{F} G_{\bar{z}} - F_{\bar{z}} \bar{G}}{F \bar{G} - \bar{F} G}, \quad b_{(F,G)} = \frac{F G_{\bar{z}} - F_{\bar{z}} G}{F \bar{G} - \bar{F} G},$$
$$A_{(F,G)} = -\frac{\bar{F} G_z - F_z \bar{G}}{F \bar{G} - \bar{F} G}, \quad B_{(F,G)} = \frac{F G_z - F_z G}{F \bar{G} - \bar{F} G}. \tag{12.12}$$

Set (for a fixed z_0)

$$W(z) = w(z) - \lambda_0 F(z) - \mu_0 G(z), \tag{12.13}$$

the constants $\lambda_0, \mu_0 \in \mathbb{R}$ being uniquely determined by the condition

$$W(z_0) = 0. \tag{12.14}$$

Hence $W(z)$ has continuous partial derivatives if and only if $w(z)$ has. Moreover, $\dot{w}(z_0)$ exists if and only if $W'(z_0)$ does, and if it does exist, then $\dot{w}(z_0) = W'(z_0)$. Therefore, by Lemma 110, if we suppose $w \in C^1(\Omega)$, the equation

$$W_{\bar{z}}(z_0) = 0 \tag{12.15}$$

is necessary and sufficient for the existence of (12.11). Now,

$$W(z) = \frac{\begin{vmatrix} w(z) & w(z_0) & \overline{w(z_0)} \\ F(z) & F(z_0) & \overline{F(z_0)} \\ G(z) & G(z_0) & \overline{G(z_0)} \end{vmatrix}}{\begin{vmatrix} F(z_0) & \overline{F(z_0)} \\ G(z_0) & \overline{G(z_0)} \end{vmatrix}} \tag{12.16}$$

so that (12.15) may be written in the form

$$\begin{vmatrix} w_{\bar{z}}(z_0) & w(z_0) & \overline{w(z_0)} \\ F_{\bar{z}}(z_0) & F(z_0) & \overline{F(z_0)} \\ G_{\bar{z}}(z_0) & G(z_0) & \overline{G(z_0)} \end{vmatrix} = 0 \tag{12.17}$$

and if (12.11) exists, then

$$\dot{w}(z_0) = \frac{\begin{vmatrix} w_z(z_0) & w(z_0) & \overline{w(z_0)} \\ F_z(z_0) & F(z_0) & \overline{F(z_0)} \\ G_z(z_0) & G(z_0) & \overline{G(z_0)} \end{vmatrix}}{\begin{vmatrix} F(z_0) & \overline{F(z_0)} \\ G(z_0) & \overline{G(z_0)} \end{vmatrix}}. \tag{12.18}$$

Equations (12.18) and (12.17) can be rewritten in the form

$$\dot{w} = w_z - A_{(F,G)}w - B_{(F,G)}\overline{w}, \tag{12.19}$$

$$w_{\bar{z}} = a_{(F,G)}w + b_{(F,G)}\overline{w}. \tag{12.20}$$

Thus we have proved the following result.

Theorem 112 ([52]). *Let (F,G) be a generating pair in some open domain Ω. Every hyperbolic function $w \in C^1(\Omega)$ admits the unique representation $w = \phi F + \psi G$ where $\phi, \psi : \Omega \subset \mathbb{D} \to \mathbb{R}$. Moreover, the (F,G)-derivative $\dot{w} = \dfrac{d_{(F,G)}w}{dz}$ of $w(z)$ exists and has the form*

$$\dot{w} = \phi_z F + \psi_z G = w_z - A_{(F,G)}w - B_{(F,G)}\overline{w} \tag{12.21}$$

if and only if

$$w_{\bar{z}} = a_{(F,G)}w + b_{(F,G)}\overline{w}. \tag{12.22}$$

The equation (12.22) can be rewritten in the form

$$\phi_{\bar{z}}F + \psi_{\bar{z}}G = 0. \tag{12.23}$$

Equation (12.22) is called a "hyperbolic Vekua equation" and any continuously differentiable solutions of this equation are called "hyperbolic (F,G)-pseudoanalytic functions". If w is hyperbolic (F,G)-pseudoanalytic, the associated function $\omega = \phi + \psi j$ is called hyperbolic (F,G)-pseudoanalytic of second kind.

One can appreciate a complete structural similarity of these first results of hyperbolic pseudoanalytic function theory with those from the elliptic theory (Section 2.2). Here we continue exploring this similarity.

Remark 113. The functions F and G are hyperbolic (F,G)-pseudoanalytic, and $\dot{F} \equiv \dot{G} \equiv 0$.

Definition 114. Let (F,G) and (F_1,G_1) be two generating pairs in Ω. (F_1,G_1) is called the successor of (F,G) and (F,G) is called the predecessor of (F_1,G_1) if

$$a_{(F_1,G_1)} = a_{(F,G)} \qquad \text{and} \qquad b_{(F_1,G_1)} = -B_{(F,G)}.$$

The importance of this definition, similarly to the elliptic case, becomes obvious from the following statement.

Theorem 115. *Let w be a hyperbolic (F, G)-pseudoanalytic function and let (F_1, G_1) be a successor of (F, G). If $\dot{w} = W \in C^1(\Omega)$ then W is a hyperbolic (F_1, G_1)-pseudoanalytic function.*

Proof. The proof in the hyperbolic case is identical to that in the elliptic case (see Theorem 11). □

Definition 116. Let (F, G) be a generating pair. Its adjoint generating pair $(F, G)^* = (F^*, G^*)$ is defined by the formulas

$$F^* = -\frac{2\overline{F}}{F\overline{G} - \overline{F}G}, \qquad G^* = \frac{2\overline{G}}{F\overline{G} - \overline{F}G}. \qquad (12.24)$$

The (F, G)-integral is defined as

$$\int_\Gamma w \, d_{(F,G)}z = F(z_1) \operatorname{Re} \int_\Gamma G^* w \, dz + G(z_1) \operatorname{Re} \int_\Gamma F^* w \, dz \qquad (12.25)$$

where Γ is a rectifiable curve leading from z_0 to z_1.

If $w = \phi F + \psi G$ is a hyperbolic (F, G)-pseudoanalytic function where ϕ and ψ are real-valued functions, then

$$\int_{z_0}^{z} \dot{w} \, d_{(F,G)}\zeta = w(z) - \phi(z_0)F(z) - \psi(z_0)G(z). \qquad (12.26)$$

This integral is path-independent and represents the (F, G)-antiderivative of $\dot{w}$. The expression $\phi(z_0)F(z) + \psi(z_0)G(z)$ in (12.26) can be seen as a *"pseudoanalytic constant"* of the generating pair (F, G) in Ω.

A continuous function $W(z)$ defined in a domain Ω will be called (F, G)-integrable if for every closed curve Γ situated in a simply connected subdomain of Ω the following equality holds:

$$\oint_\Gamma W d_{(F,G)}z = 0. \qquad (12.27)$$

Theorem 117. *Let W be a hyperbolic (F, G)-pseudoanalytic function. Then W is (F, G)-integrable.*

Proof. It will suffice to show that if Ω is a regular domain and Γ lies within the domain of definition of W, then

$$\int_\Gamma W d_{(F,G)}z \qquad (12.28)$$

is zero.

From the definitions (12.12) and (12.24) we find

$$a_{(F^*,G^*)} = -a_{(F,G)}, \quad A_{(F^*,G^*)} = -A_{(F,G)},$$
$$b_{(F^*,G^*)} = -\overline{B_{(F,G)}}, \quad B_{(F^*,G^*)} = -\overline{b_{(F,G)}}. \tag{12.29}$$

Hence we obtain

$$F_{\bar{z}}^* = -aF^* - \overline{B}\,\overline{F^*}, \quad G_{\bar{z}}^* = -aG^* - \overline{B}\,\overline{G^*}, \tag{12.30}$$

and by hypothesis

$$W_{\bar{z}} = aW - B\overline{W}, \tag{12.31}$$

where a, b, A and B are the characteristic coefficients of (F,G).

Let us now use the definition (12.25) to evaluate (12.28). By using the hyperbolic Green's theorem (see [52]), we obtain

$$\int_\Gamma G^* W \mathrm{d}z = 2\mathrm{j} \int\int_\Omega \partial_{\bar{z}}(G^*W)\mathrm{d}x\mathrm{d}t$$
$$= 2\mathrm{j} \int\int_\Omega \left(-aG^*W - \overline{B}\,\overline{G^*}W + G^*aW - G^*B\overline{W}\right)\mathrm{d}x\mathrm{d}t$$
$$= -4\mathrm{j} \int\int_\Omega \mathrm{Re}\left(G^*B\overline{W}\right)\mathrm{d}x\mathrm{d}t$$

which is a purely imaginary number. The same argument shows that $\int_\Gamma F^* W \mathrm{d}z$ is a pure imaginary number. Hence by Definition (12.25) we find that (12.28) is zero. □

12.3   Generating sequences

Definition 118. A sequence of generating pairs $\{(F_m, G_m)\}$ with $m \in \mathbb{Z}$, is called a generating sequence if (F_{m+1}, G_{m+1}) is a successor of (F_m, G_m). If $(F_0, G_0) = (F,G)$, we say that (F,G) is embedded in $\{(F_m, G_m)\}$.

Definition 119. A generating sequence $\{(F_m, G_m)\}$ is said to have period $\mu > 0$ if $(F_{m+\mu}, G_{m+\mu})$ is equivalent to (F_m, G_m), that is their characteristic coefficients coincide.

Let w be a hyperbolic (F,G)-pseudoanalytic function. Using a generating sequence in which (F,G) is embedded we can define the higher derivatives of w by the recursion formula

$$w^{[0]} = w; \quad w^{[m+1]} = \frac{\mathrm{d}_{(F_m,G_m)}w^{[m]}}{\mathrm{d}z}, \quad m = 0,1,2,\ldots.$$

Definition 120. The formal power $Z_m^{(0)}(a, z_0; z)$ with center at $z_0 \in \Omega$, coefficient a and exponent 0 is defined as the linear combination of the generators F_m, G_m with real constant coefficients λ, μ chosen so that $\lambda F_m(z_0) + \mu G_m(z_0) = a$. The formal powers with exponents $n = 1, 2, \ldots$ are defined by the recursion formula

$$Z_m^{(n)}(a, z_0; z) = n \int_{z_0}^{z} Z_{m+1}^{(n-1)}(a, z_0; \zeta) \mathrm{d}_{(F_m, G_m)} \zeta. \qquad (12.32)$$

This definition implies the following properties similar to those in the elliptic case (cf. Section 4.1).

1. $Z_m^{(n)}(a, z_0; z)$ is a (F_m, G_m)-hyperbolic pseudoanalytic function of z.

2. If a' and a'' are real constants, then $Z_m^{(n)}(a' + \mathrm{j}a'', z_0; z) = a' Z_m^{(n)}(1, z_0; z) + a'' Z_m^{(n)}(\mathrm{j}, z_0; z)$.

3. The formal powers satisfy the differential relations

$$\frac{\mathrm{d}_{(F_m, G_m)} Z_m^{(n)}(a, z_0; z)}{dz} = n Z_{m+1}^{(n-1)}(a, z_0; z).$$

4. The asymptotic formulas

$$Z_m^{(n)}(a, z_0; z) \sim a(z - z_0)^n, \quad z \to z_0$$

hold.

Chapter 13

Relationship between Hyperbolic Pseudoanalytic Functions and Solutions of the Klein-Gordon Equation

13.1 Factorization of the Klein-Gordon equation

Consider the $(1+1)$-dimensional Klein-Gordon equation

$$\left(\Box - \nu(x,t)\right)\varphi(x,t) = 0 \tag{13.1}$$

in some domain $\Omega \subset \mathbb{R}^2$, where

$$\Box := \frac{\partial^2}{\partial x^2} - \frac{\partial^2}{\partial t^2}, \quad \nu \quad \text{and} \quad \varphi$$

are real-valued functions. We assume that φ is a twice-continuously differentiable function.

In analogy to the factorization of the stationary two-dimensional Schrödinger equation (Section 3.1) it is possible to factorize the Klein-Gordon equation with potential. By C we denote the hyperbolic conjugation operator $(Cj = -j)$.

Theorem 121 ([80]). *Let f be a positive particular solution of (13.1) in Ω. Then for any real-valued function $\varphi \in C^2(\Omega)$ the following equalities hold:*

$$\begin{aligned}
(\Box - \nu)\varphi &= 4\left(\partial_{\bar{z}} + \frac{f_z}{f}C\right)\left(\partial_z - \frac{f_z}{f}C\right)\varphi \\
&= 4\left(\partial_z + \frac{f_{\bar{z}}}{f}C\right)\left(\partial_{\bar{z}} - \frac{f_{\bar{z}}}{f}C\right)\varphi.
\end{aligned} \tag{13.2}$$

Proof. Consider

$$\left(\partial_{\bar{z}} + \frac{f_z}{f}C\right)\left(\partial_z - \frac{f_z}{f}C\right)\varphi = \partial_{\bar{z}}\partial_z\varphi - \frac{|f_z|^2}{f^2}\varphi - \partial_{\bar{z}}\left(\frac{f_z}{f}\right)\varphi$$

$$= \frac{1}{4}\left(\Box\varphi - \frac{\Box f}{f}\varphi\right) = \frac{1}{4}(\Box - \nu)\varphi. \tag{13.3}$$

Thus we have the first equality in (13.2). Now application of C to both sides of (13.3) gives us the second equality in (13.2). $\Box$

Note that the operator $\partial_z - \frac{f_z}{f}I$, where I is the identity operator, can be represented in the form

$$P = \partial_z - \frac{f_z}{f}I = f\partial_z f^{-1}I. \tag{13.4}$$

Let us introduce the notation $P := f\partial_z f^{-1}I$. From Theorem 121, if f is a positive solution of (13.1), the operator P transforms real-valued solutions of (13.1) into solutions of the hyperbolic Vekua equation

$$\left(\partial_{\bar{z}} + \frac{f_z}{f}C\right)w = 0. \tag{13.5}$$

The operator ∂_z applied to a real-valued function φ can be regarded as a kind of gradient. If we have $\partial_z\varphi = \Phi$ in a convex hyperbolic domain, where $\Phi = \Phi_1 + j\Phi_2$ is a given hyperbolic-valued function such that its real part Φ_1 and imaginary part Φ_2 satisfy

$$\partial_t\Phi_1 - \partial_x\Phi_2 = 0, \tag{13.6}$$

then we can construct φ up to an arbitrary real constant c. Indeed, we have

$$\varphi(x,t) = 2\left(\int_{x_0}^x \Phi_1(\eta,t)d\eta + \int_{t_0}^t \Phi_2(x_0,\xi)d\xi\right) + c \tag{13.7}$$

where (x_0, t_0) is an arbitrary fixed point in the domain of interest. We will denote the integral operator in (13.7) by A_h:

$$A_h[\Phi](x,t) = 2\left(\int_{x_0}^x \Phi_1(\eta,t)d\eta + \int_{t_0}^t \Phi_2(x_0,\xi)d\xi\right) + c. \tag{13.8}$$

Thus if Φ satisfies (13.6), there exists a family of real-valued functions φ such that $\partial_z\varphi = \Phi$, given by $\varphi = A_h[\Phi]$.

In a similar way we define the operator $\bar{A}_h$ corresponding to $\partial_{\bar{z}}$. The family of real-valued functions φ such that $\partial_{\bar{z}}\varphi = \Phi$, where $\varphi = \overline{A}_h[\Phi]$, can be constructed as

$$\bar{A}_h[\Phi](x,t) = 2\left(\int_{x_0}^x \Phi_1(\eta,t)d\eta - \int_{t_0}^t \Phi_2(x_0,\xi)d\xi\right) + c, \tag{13.9}$$

when

$$\partial_t \Phi_1 + \partial_x \Phi_2 = 0. \tag{13.10}$$

Note that both definitions A_h and $\overline{A}_h$ are easily extended to any simply connected domain.

Consider the operator $S = f A_h f^{-1} I$ applicable to any hyperbolic-valued function w such that $\Phi = f^{-1} w$ satisfies condition (13.6). Then it is clear that for such w we have that $PSw = w$.

Theorem 122 ([80]). *Let f be a positive particular solution of* (13.1) *and w be a solution of* (13.5). *Then the real-valued function $g = Sw$ is a solution of* (13.1).

Proof. First, let us verify that the function $\Phi = w/f$ satisfies condition (13.6). Let $w = u + \mathrm{j}v$. We find

$$\partial_t \Phi_1 - \partial_x \Phi_2 = f^{-1} \cdot \left((u_t - v_x) - \left(\frac{f_t}{f} u - \frac{f_x}{f} v \right) \right). \tag{13.11}$$

The equation (13.5) is equivalent to the system

$$u_x - v_t = -\frac{f_x}{f} u + \frac{f_t}{f} v, \quad u_t - v_x = \frac{f_t}{f} u - \frac{f_x}{f} v \tag{13.12}$$

and we find that the expression (13.11) is zero. Hence, the function $\Phi = w/f$ satisfies (13.6) and we have $PSw = w$.

Let $Q = (\partial_{\bar{z}} + \frac{f_{\bar{z}}}{f} C)$ such that $QP = \frac{1}{4}(\square - \nu)$ from Theorem 121. We obtain

$$PSw = w \quad \Rightarrow \quad QPSw = Qw = 0 \quad \Rightarrow \quad \frac{1}{4}(\square - \nu)Sw = 0. \tag{13.13}$$

$$\square$$

13.2 The main hyperbolic Vekua equation

The Vekua equation (13.5) is closely related to another Vekua equation given by

$$\left(\partial_{\bar{z}} - \frac{f_{\bar{z}}}{f} C \right) W = 0. \tag{13.14}$$

Indeed, one can observe that the pair of functions

$$F = f \quad \text{and} \quad G = \frac{\mathrm{j}}{f} \tag{13.15}$$

is a generating pair for (13.14). The associated characteristic coefficients are then given by

$$A_{(F,G)} = 0, \quad B_{(F,G)} = \frac{f_z}{f}, \quad a_{(F,G)} = 0, \quad b_{(F,G)} = \frac{f_{\bar{z}}}{f}, \tag{13.16}$$

and the (F, G)-derivative according to (12.21) is defined as

$$\dot{W} = W_z - \frac{f_z}{f}\overline{W} = \left(\partial_z - \frac{f_z}{f}C\right)W. \tag{13.17}$$

From Definition 114 and Theorem 115, if we compare $B_{(F,G)}$ with the coefficient in (13.5) we obtain the following statement.

Theorem 123. *If $W \in C^1(\Omega)$ is a solution of (13.14), then its (F, G)-derivative $\dot{W} = w$ is a solution of (13.5) on Ω.*

Let us now consider the (F, G)-antiderivative. Taking into account that $F^* \doteq jf$ and $G^* = 1/f$, we find

$$\int_{z_0}^{z} w(\zeta)\mathrm{d}_{(F,G)}\zeta = f(z)\mathrm{Re}\int_{z_0}^{z} \frac{w(\zeta)}{f(\zeta)}\mathrm{d}\zeta - \frac{j}{f(z)}\mathrm{Re}\int_{z_0}^{z} jf(\zeta)w(\zeta)\mathrm{d}\zeta$$

$$= f(z)\mathrm{Re}\int_{z_0}^{z} \frac{w(\zeta)}{f(\zeta)}\mathrm{d}\zeta + \frac{j}{f(z)}\mathrm{Im}\int_{z_0}^{z} f(\zeta)w(\zeta)\mathrm{d}\zeta \tag{13.18}$$

and we obtain the following statement.

Theorem 124. *If w is a solution of (13.5), then the function*

$$W(z) = \int_{z_0}^{z} w(\zeta)\mathrm{d}_{(F,G)}\zeta \tag{13.19}$$

is a solution of (13.14).

Lemma 125. *Let b be a hyperbolic function such that b_z is a real-valued function, and let $W = u + jv$ be a solution of the equation*

$$W_{\bar{z}} = b\overline{W}. \tag{13.20}$$

Then u is a solution of the equation

$$\frac{1}{4}\Box u = (b\bar{b} + b_z)u \tag{13.21}$$

and v is a solution of the equation

$$\frac{1}{4}\Box v = (b\bar{b} - b_z)v. \tag{13.22}$$

Proof. We observe that under the conjugation the equation $W_{\bar{z}} = b\overline{W}$ is equivalent to $\partial_z(u - jv) = \bar{b}(u + jv)$. Then we obtain

$$\frac{1}{4}\Box(u + jv) = \partial_z\partial_{\bar{z}}(u + jv)$$

$$= b_z(u - jv) + b\partial_z(u - jv) \tag{13.23}$$

$$= b_z(u - jv) + b\bar{b}(u + jv)$$

and by considering the real and imaginary parts of this expression we complete the proof. $\square$

Theorem 126. *Let W be a solution of (13.14). Then $u = \operatorname{Re} W$ is a solution of (13.1) and $v = \operatorname{Im} W$ is a solution of the equation*

$$\left(\square - \eta\right)v = 0, \quad \text{where} \quad \eta = -\nu + 8\frac{|f_z|^2}{f^2}. \tag{13.24}$$

Proof. Let us first show that if $b = \dfrac{f_{\bar{z}}}{f}$, then b_z is a real-valued function:

$$b_z = \frac{(\partial_z f_{\bar{z}})f - f_{\bar{z}}f_z}{f^2} = \frac{1}{4}\frac{\square f}{f} - \frac{|f_z|^2}{f^2} = \frac{1}{4}\nu - \frac{|f_z|^2}{f^2} \in \mathbb{R}. \tag{13.25}$$

We can easily calculate that $4(b\bar{b} + b_z) = \nu$ and $4(b\bar{b} - b_z) = \eta$ such that according to Lemma 125 we find $(\square - \nu)u = 0$ and $(\square - \eta)v = 0$. □

Remark 127. If we consider the case $\nu = 0$ in (13.1), then we obtain the one-dimensional wave equation $\square\varphi = 0$ with the well-known general solution $\varphi = F(x+t) + G(x-t)$, where F and G are two arbitrary real-valued functions of one variable. In this case, the potential η is then given by $\eta = 8\dfrac{F'G'}{(F+G)^2}$.

Theorem 128. *Let u be a solution of (13.1). Then the function $v \in \ker(\square - \eta)$, such that $W = u + \mathrm{j}v$ is a solution of (13.14), is constructed according to the formula*

$$v = -f^{-1}\overline{A}_h\left[\mathrm{j}f^2\partial_{\bar{z}}(f^{-1}u)\right]. \tag{13.26}$$

It is unique up to an additive term cf^{-1} where c is an arbitrary real constant.

Let v be a solution of (13.24). Then the function $u \in \ker(\square - \nu)$, such that $W = u + \mathrm{j}v$ is a solution of (13.14), can be constructed as

$$u = -f\overline{A}_h\left[\mathrm{j}f^{-2}\partial_{\bar{z}}(fv)\right], \tag{13.27}$$

up to an additive term cf.

Proof. Consider $W = \phi f + \mathrm{j}\psi/f$ to be a solution of the Vekua equation (13.14). Then this equation can be rewritten in the form

$$\psi_{\bar{z}} = -\mathrm{j}f^2\phi_{\bar{z}}$$
$$= \frac{f^2}{2}(\phi_t - \mathrm{j}\phi_x). \tag{13.28}$$

Taking into account that $\phi = u/f$, $(\square - \nu)u = 0$ and $(\square - \nu)f = 0$, we can verify that

$$\partial_t\left(\frac{f^2}{2}\phi_t\right) + \partial_x\left(\frac{f^2}{2}\phi_x\right) = 0, \tag{13.29}$$

such that we can use (13.9) and ψ is given by $\psi = -\overline{A}_h\left[\mathrm{j}f^2\phi_{\bar{z}}\right]$. Now, since $v = \operatorname{Im} W = \psi/f$ we find $v = -f^{-1}\overline{A}_h\left[\mathrm{j}f^2\partial_{\bar{z}}(f^{-1}u)\right]$. The function v is a solution of (13.24) due to Theorem 126. Note that as the operator $\overline{A}_h$ reconstructs the scalar function up to an arbitrary real constant, the function v in formula (13.26) is uniquely determined up to an additive term cf^{-1} where c is an arbitrary real constant.

Equation (13.27) is proved in a similar way. $\square$

Example 129. Let us illustrate the last theorem by a simple example. Considering $f(x,t) = xt = \frac{1}{4}\big((x+t)^2 - (x-t)^2\big)$ and $u(x,t) = 1$ to be two particular solutions of the wave equation in the subdomain $0 < x < t < \infty$, then $v = -f^{-1}\overline{A}_h\big[\mathrm{j}f^2\partial_{\bar{z}}(f^{-1}u)\big] \in \ker\,(\square - \eta)$, where

$$\eta(x,t) = 8\frac{|f_z|^2}{f^2} = 2\frac{t^2 - x^2}{x^2 t^2}. \tag{13.30}$$

Explicitly, the solution v is given by

$$v(x,t) = \frac{x^2 + t^2}{2xt}. \tag{13.31}$$

13.3   Generating sequence for the main hyperbolic Vekua equation

The first step in the construction of a generating sequence for the main hyperbolic Vekua equation (13.14) is the construction of a generating pair for the equation (13.5) which, as was shown previously, is a successor of the main Vekua equation. For this, one of the possibilities consists in constructing another pair of solutions of (13.14). Then their (F,G)-derivatives will give us solutions of (13.5).

Consider the main Vekua equation (13.14) which is equivalent to the equation

$$\phi_{\bar{z}}F + \psi_{\bar{z}}G = 0, \tag{13.32}$$

where $W = \phi F + \psi G$, $F = f$ and $G = \mathrm{j}/f$. Equation (13.32) can be rewritten explicitly as the system of partial differential equations

$$\phi_x f^2 - \psi_t = 0,$$
$$\psi_x - \phi_t f^2 = 0. \tag{13.33}$$

Let us suppose that f and ϕ are functions of some real variable $\rho = \rho(x,t)$, i.e., $f = f(\rho)$ and $\phi = \phi(\rho)$. System (13.33) then becomes

$$\psi_x = \phi' \rho_t f^2,$$
$$\psi_t = \phi' \rho_x f^2. \tag{13.34}$$

The compatibility condition for this system implies

$$\partial_x \left(\phi' \rho_x f^2 \right) - \partial_t \left(\phi' \rho_t f^2 \right) = 0, \tag{13.35}$$

which is equivalent to the equation

$$\phi'' + \left(\frac{\Box \rho}{4|\rho_z|^2} + 2\frac{f'}{f} \right) \phi' = 0, \tag{13.36}$$

for $|\rho_z|^2 \neq 0$. We assume now that $\frac{\Box \rho}{4|\rho_z|^2}$ is a function of ρ, i.e.,

$$\frac{\Box \rho}{4|\rho_z|^2} = s(\rho). \tag{13.37}$$

Hence, under this hypothesis, we can integrate (13.36) and obtain

$$\phi'(\rho) = \frac{e^{-S(\rho)}}{f^2}, \tag{13.38}$$

where $S(\rho) = \int_{\rho_0}^{\rho} s(\sigma) \, d\sigma$.

We can now integrate (13.38) and (13.34) to obtain a solution $W = \phi F + \psi G$ of (13.14). However, since we are interested in finding a solution of (13.5), i.e., the (F, G)-derivative $\dot{W}$, we need ϕ_z and ψ_z which are given explicitly by

$$\phi_z = \frac{e^{-S} \rho_z}{f^2}, $$
$$\psi_z = \frac{j}{2} e^{-S} \rho_z. \tag{13.39}$$

Thus, a solution $w_1 = \phi_z F + \psi_z G$ of (13.5) is given by

$$w_1 = \frac{3}{2} e^{-S} \frac{\rho_z}{f}. \tag{13.40}$$

In much the same way we can construct another solution of (13.5) looking for $\psi = \psi(\rho)$. The system (13.33) then becomes

$$\phi_x = \frac{\psi' \rho_t}{f^2}, $$
$$\phi_t = \frac{\psi' \rho_x}{f^2} \tag{13.41}$$

and $\psi'(\rho) = f^2 e^{-S(\rho)}$. Calculating ϕ_z and ψ_z we find

$$\phi_z = \frac{j}{2} e^{-S} \rho_z, $$
$$\psi_z = f^2 e^{-S} \rho_z, \tag{13.42}$$

which gives us another solution w_2 of (13.5):

$$w_2 = \frac{3}{2}\mathrm{j}\,\mathrm{e}^{-S}\rho_z f. \tag{13.43}$$

Hence, for the function $\Phi = \mathrm{j}\,\mathrm{e}^{-S}\rho_z \neq 0$ we have found a generating pair for the Vekua equation (13.5) given by (eliminating the constant $\frac{3}{2}$ in w_1 and w_2):

$$\begin{aligned}(F_1, G_1) &= \left(\mathrm{j}\,\mathrm{e}^{-S}\rho_z f,\ \mathrm{j}\,\mathrm{e}^{-S}\rho_z \frac{\mathrm{j}}{f}\right) \\ &= (\Phi F, \Phi G).\end{aligned} \tag{13.44}$$

Indeed, we have

$$\mathrm{Im}\big(\overline{F_1}G_1\big) = \mathrm{Im}\big(|\Phi|^2 \overline{F}G\big) = -\mathrm{e}^{-S}|\rho_z|^2 \neq 0. \tag{13.45}$$

The following step is to construct the generating pair (F_2, G_2). For this we should find two other solutions of (13.5), equivalent to $\phi_{\bar{z}}F_1 + \psi_{\bar{z}}G_1 = 0$. Then to obtain (F_2, G_2) we calculate the (F_1, G_1)-derivative of these solutions. Using the same assumptions and the same method as in the previous case, we obtain

$$(F_2, G_2) = (\Phi^2 F, \Phi^2 G). \tag{13.46}$$

The generalization of results (13.44) and (13.46) is given in the next theorem which allows us to obtain a generating sequence wherein the generating pair (F, G) of (13.14) is embedded. Let us note that under the assumption (13.37) the function Φ is a "hyperbolic analytic function", i.e., $\Phi_{\bar{z}} = 0$. Indeed, we have

$$\begin{aligned}\Phi_{\bar{z}} &= \mathrm{j}\left((\partial_{\bar{z}}\mathrm{e}^{-S})\rho_z + \frac{1}{4}\mathrm{e}^{-S}\Box\rho\right) \\ &= -\frac{1}{4}\mathrm{j}\,\mathrm{e}^{-S}\left(4s|\rho_z|^2 - \Box\rho\right) = 0.\end{aligned}$$

Theorem 130 ([80]). *Let f be a nonvanishing solution of* (13.1) *such that $f = f(\rho)$, $\rho = \rho(x, t)$, and $\frac{\Box\rho}{4|\rho_z|^2}$ is a function of ρ denoted by $s(\rho)$. Let also the function Φ such that $\Phi = \mathrm{j}\,\mathrm{e}^{-S(\rho)}\rho_z \neq 0$, where $S(\rho) = \int_{\rho_0}^{\rho} s(\sigma)\mathrm{d}\sigma$. Then the generating pair (F, G) with $F = f$ and $G = \mathrm{j}/f$ is embedded in the generating sequence (F_m, G_m) where $F_m = \Phi^m F$, $G_m = \Phi^m G$ and $m \in \mathbb{Z}$.*

Proof. First, let us show that (F_m, G_m) is a generating pair in $\mathbb{Z}$. Indeed, we find

$$\mathrm{Im}\big(\overline{F_m}G_m\big) = \mathrm{Im}\big(|\Phi|^{2m}\overline{F}G\big) = (-1)^m \mathrm{e}^{-2mS}|\rho_z|^{2m} \neq 0.$$

To complete the proof, we need to show that $\big\{(F_m, G_m)\big\}$ forms a generating sequence, i.e., (F_m, G_m) is a successor of (F_{m-1}, G_{m-1}):

$$a_{(F_m, G_m)} = a_{(F_{m-1}, G_{m-1})} \quad \text{and} \quad b_{(F_m, G_m)} = -B_{(F_{m-1}, G_{m-1})}. \tag{13.47}$$

The coefficients $a_{(F_m, G_m)}$, $b_{(F_m, G_m)}$ and $B_{(F_m, G_m)}$ can be calculated in terms of $a_{(F,G)}$, $b_{(F,G)}$ and $B_{(F,G)}$, respectively, by taking into account that $\Phi_{\bar{z}} = 0$. We obtain

$$a_{(F_m, G_m)} = |\Phi|^{2m} a_{(F,G)} = 0, \quad b_{(F_m, G_m)} = \left(\frac{\Phi}{\overline{\Phi}}\right)^m b_{(F,G)},$$
$$B_{(F_m, G_m)} = \left(\frac{\Phi}{\overline{\Phi}}\right)^m B_{(F,G)}.$$
(13.48)

Therefore, the first equality in (13.47) is verified. Taking into account (13.16) and (13.48), the second equality in (13.47) is reduced to

$$\overline{\Phi} f_z + \Phi f_{\bar{z}} = 0 \quad \Leftrightarrow \quad f'(\overline{\Phi} \rho_z + \Phi \rho_z) = 0.$$
(13.49)

Since $\Phi = \mathrm{j}\, e^{-S(\rho)} \rho_z$ it is easy to observe that (13.49) is valid. $\square$

This last theorem allows us to calculate the generating sequence (F_m, G_m) for a large class of potentials $\nu(x, t)$ in the Klein-Gordon equation (13.1). The importance of this result appears in the following theorem.

Theorem 131. *Let f be a particular solution of (13.1) and let (F, G) be the generating pair in some open domain Ω with $F = f$ and $G = \mathrm{j}/f$. Then*

$$\mathrm{Re}\, Z^{(n)}(a, z_0; z), \quad n = 0, 1, 2, \ldots$$

are solutions of the Klein-Gordon equation (13.1).

Proof. From property 1 of Definition 120 we see that $Z^{(n)}(a, z_0; z)$ is a hyperbolic (F, G)-pseudoanalytic function. Hence $Z^{(n)}(a, z_0; z)$ satisfies (13.14) and the real parts are solutions of (13.1) from Theorem 126. $\square$

Example 132. As an example of this theorem, we consider the Klein-Gordon equation (13.1) with the potential $\nu(x, t) = t^2 - x^2$ in the "time-like" subdomain $0 < x < t < \infty$. A particular solution of this equation is given by $f(\rho) = \exp(\rho^2)$, where we have defined $\rho(x, t) = \sqrt{xt}$. In this case the function $\frac{\Box \rho}{4|\rho_z|^2}$ is a function of ρ given by $s(\rho) = -1/\rho$, with $S(\rho) = -\ln \rho$ and $\Phi = \frac{z}{4} \neq 0$. Let us construct the first formal powers $Z^{(n)}(1, 4\mathrm{j}; z)$ and $Z^{(n)}(\mathrm{j}, 4\mathrm{j}; z)$. By Definition 120 we obtain

$$
\begin{aligned}
Z^{(0)}(1, 4\mathrm{j}; 4\mathrm{j}) &= 1, & Z^{(0)}(\mathrm{j}, 4\mathrm{j}; 4\mathrm{j}) &= \mathrm{j}, \\
&= \lambda_1 F(4\mathrm{j}) + \mu_1 G(4\mathrm{j}), & &= \lambda_2 F(4\mathrm{j}) + \mu_2 G(4\mathrm{j}), \\
&= \lambda_1 + \mathrm{j}\mu_1, & &= \lambda_2 + \mathrm{j}\mu_2,
\end{aligned}
$$

such that $\lambda_1 = \mu_2 = 1$ and $\mu_1 = \lambda_2 = 0$. Hence, we obtain

$$
\begin{aligned}
Z^{(0)}(1, 4\mathrm{j}; z) &= \lambda_1 F(z) + \mu_1 G(z), & Z^{(0)}(\mathrm{j}, 4\mathrm{j}; z) &= \lambda_2 F(z) + \mu_2 G(z) \\
&= e^{xt}, & &= \mathrm{j}e^{-xt}.
\end{aligned}
$$

Now, from the formula (12.32), if we want to construct

$$Z^{(1)}(1, 4\mathrm{j}; z) \qquad \text{and} \qquad Z^{(1)}(\mathrm{j}, 4\mathrm{j}; z)$$

we need to calculate first $Z_1^{(0)}(1, 4\mathrm{j}; z)$ and $Z_1^{(0)}(\mathrm{j}, 4\mathrm{j}; z)$. Note that the generating pair (F_1, G_1) is given by

$$F_1 = \frac{1}{4} z\mathrm{e}^{xt} \quad \text{and} \quad G_1 = \frac{\mathrm{j}}{4} z\mathrm{e}^{-xt}.$$

Hence, from Definition 120 we obtain

$$\begin{aligned}
Z_1^{(0)}(1, 4\mathrm{j}; 4\mathrm{j}) &= 1, \\
&= \lambda_3 F_1(4\mathrm{j}) + \mu_3 G_1(4\mathrm{j}), \\
&= \lambda_3 \mathrm{j} + \mu_3,
\end{aligned}$$

$$\begin{aligned}
Z_1^{(0)}(\mathrm{j}, 4\mathrm{j}; 4\mathrm{j}) &= \mathrm{j}, \\
&= \lambda_4 F_1(4\mathrm{j}) + \mu_4 G_1(4\mathrm{j}), \\
&= \lambda_4 \mathrm{j} + \mu_4,
\end{aligned}$$

which implies that $\mu_3 = \lambda_4 = 1$ and $\lambda_3 = \mu_4 = 0$ and

$$Z_1^{(0)}(1, 4\mathrm{j}; z) = \lambda_3 F_1(z) + \mu_3 G_1(z), \qquad Z_1^{(0)}(\mathrm{j}, 4\mathrm{j}; z) = \lambda_4 F_1(z) + \mu_4 G_1(z)$$

$$= \frac{\mathrm{j}}{4} z\mathrm{e}^{-xt} \qquad\qquad\qquad\qquad\qquad = \frac{1}{4} z\mathrm{e}^{xt}.$$

From the definition (12.32), we have

$$Z^{(1)}(a, 4\mathrm{j}; z) = \int_0^z Z_1^{(0)}(a, 4\mathrm{j}; \zeta) \mathrm{d}_{(F,G)}\zeta$$

and (12.24) gives $F^* = \mathrm{j}f$ and $G^* = 1/f$. Now using (12.25) we find

$$Z^{(1)}(1, 4\mathrm{j}; z) = \frac{1}{4}\left(\mathrm{e}^{xt}\mathrm{Re}\int_0^z \mathrm{j}\mathrm{e}^{-2x't'}\zeta d\zeta + \mathrm{j}\mathrm{e}^{-xt}\mathrm{Re}\int_0^z \zeta d\zeta\right),$$

$$Z^{(1)}(\mathrm{j}, 4\mathrm{j}; z) = \frac{1}{4}\left(\mathrm{e}^{xt}\mathrm{Re}\int_0^z \zeta d\zeta + \mathrm{j}\mathrm{e}^{-xt}\mathrm{Re}\int_0^z \mathrm{j}\mathrm{e}^{2x't'}\zeta d\zeta\right),$$

where $\zeta = x' + \mathrm{j}t'$. Evaluating these integrals, we obtain

$$\mathrm{Re}\int_0^z \zeta d\zeta = \mathrm{Re}\int_0^1 \epsilon(x + \mathrm{j}t)(x + \mathrm{j}t)\mathrm{d}\epsilon = \frac{x^2 + t^2}{2}$$

and

$$\mathrm{Re}\int_0^z \mathrm{j}\mathrm{e}^{\pm 2x't'}\zeta d\zeta = \mathrm{Re}\int_0^1 \mathrm{j}(x + \mathrm{j}t)^2 \epsilon\mathrm{e}^{\pm 2\epsilon^2 xt}\mathrm{d}\epsilon = \mathrm{e}^{\pm xt}\sinh(xt),$$

such that

$$Z^{(1)}(1, 4\mathrm{j}; z) = \frac{1}{4}\left(\sinh(xt) + \mathrm{j}\frac{x^2 + t^2}{2}\mathrm{e}^{-xt}\right),$$

$$Z^{(1)}(\mathrm{j}, 4\mathrm{j}; z) = \frac{1}{4}\left(\frac{x^2 + t^2}{2}\mathrm{e}^{xt} + \mathrm{j}\sinh(xt)\right).$$

Now, according to (12.32), if we want to find $Z^{(2)}(1, 4\mathrm{j}; z)$ and $Z^{(2)}(\mathrm{j}, 4\mathrm{j}; z)$ we need to calculate first $Z_1^{(1)}(1, 4\mathrm{j}; z)$ and $Z_1^{(1)}(\mathrm{j}, 4\mathrm{j}; z)$; those are themselves obtained from $Z_2^{(0)}(1, 4\mathrm{j}; z)$ and $Z_2^{(0)}(\mathrm{j}, 4\mathrm{j}; z)$.

The generating pair (F_2, G_2) is given by

$$F_2 = \left(\frac{z}{4}\right)^2 \mathrm{e}^{xt} \quad \text{and} \quad G_2 = \mathrm{j}\left(\frac{z}{4}\right)^2 \mathrm{e}^{-xt},$$

which allows us to calculate $Z_2^{(0)}(1, 4\mathrm{j}; z)$ and $Z_2^{(0)}(\mathrm{j}, 4\mathrm{j}; z)$. We find

$$Z_2^{(0)}(1, 4\mathrm{j}; z) = \left(\frac{z}{4}\right)^2 \mathrm{e}^{xt}, \quad Z_2^{(0)}(\mathrm{j}, 4\mathrm{j}; z) = \mathrm{j}\left(\frac{z}{4}\right)^2 \mathrm{e}^{-xt}.$$

To obtain $Z_1^{(1)}(1, 4\mathrm{j}; z)$ and $Z_1^{(1)}(\mathrm{j}, 4\mathrm{j}; z)$, we need the adjoint generating pair of (F_1, G_1), i.e.,

$$F_1^* = 4\mathrm{j}\frac{\mathrm{e}^{xt}}{z}, \quad G_1^* = 4\frac{\mathrm{e}^{-xt}}{z}.$$

Using (12.32), we find

$$Z_1^{(1)}(1, 4\mathrm{j}; z) = \frac{1}{16}(x + \mathrm{j}t)\left(\frac{x^2 + t^2}{2}\mathrm{e}^{xt} + \mathrm{j}\sinh(xt)\right),$$

$$Z_1^{(1)}(\mathrm{j}, 4\mathrm{j}; z) = \frac{1}{16}(x + \mathrm{j}t)\left(\sinh(xt) + \mathrm{j}\frac{x^2 + t^2}{2}\mathrm{e}^{-xt}\right).$$

Finally, by considering again (12.32), we obtain

$$Z^{(2)}(1, 4\mathrm{j}; z) = \frac{\mathrm{e}^{xt}}{64}\left[(x^2 + t^2)^2 + 4xt + 2\big(\cosh(2xt) - \sinh(2xt) - 1\big)\right]$$
$$+ \frac{\mathrm{j}}{64}\frac{x^2 + t^2}{xt}\left[\mathrm{e}^{xt}\big(4xt\sinh(2xt) - 1\big)\right.$$
$$\left. + 2\mathrm{e}^{-xt}\cosh(xt)\big(\cosh(xt) + \sinh(xt)\big) - 1\right],$$

$$Z^{(2)}(\mathrm{j}, 4\mathrm{j}; z) = -\frac{1}{64}\frac{x^2 + t^2}{xt}\left[\mathrm{e}^{xt}\big(\sinh(2xt) - \cosh(2xt)\big) - 4xt\sinh(xt) + \mathrm{e}^{-xt}\right]$$
$$- \frac{\mathrm{j}}{64}\mathrm{e}^{-xt}\left[(x^2 + t^2)^2 + 4\cosh(xt)\big(\sinh(xt)\right.$$
$$\left. + \cosh(xt)\big) - 4\big(xt + 1\big)\right].$$

According to Theorem 131, we can verify that the real parts of $Z^{(n)}(1, 4\mathrm{j}; z)$ and $Z^{(n)}(\mathrm{j}, 4\mathrm{j}; z)$ with $n = 0, 1, 2, \ldots$ are solutions of the Klein-Gordon equation with potential $\nu(x, t) = t^2 - x^2$.

Example 133. We are now considering the Klein-Gordon equation (13.1) with the potential

$$\nu(x, t) = \frac{1}{4}\left(\frac{1}{(t+1)^2} - \frac{1}{(x+1)^2}\right) \tag{13.50}$$

on the time-like subdomain $0 < x < t < \infty$. A particular solution of this equation is given by $f(x, t) = \sqrt{(x+1)(t+1)}$. We write $\rho = (x+1)(t+1)$. In this case it is easy to see that the function $\frac{\Box\rho}{4|\rho_z|^2}$ is zero, therefore a function of ρ. We obtain $\Phi = \frac{z}{2} + \mathrm{e}_1$, where e_1 is the idempotent constant of (11.5). Let us calculate the first formal powers $Z^{(n)}(1, t_0\mathrm{j}; z)$ and $Z^{(n)}(\mathrm{j}, t_0\mathrm{j}; z)$, where $t_0 > 0$. We find

$$Z^{(0)}(1, t_0\mathrm{j}; z) = \alpha^{-1}\sqrt{(x+1)(t+1)},$$

$$Z^{(0)}(\mathrm{j}, t_0\mathrm{j}; z) = \frac{\mathrm{j}\alpha}{\sqrt{(x+1)(t+1)}}.$$

where $\alpha = \sqrt{t_0 + 1}$.

From (12.32), in order to construct $Z^{(1)}(1, t_0\mathrm{j}; z)$ and $Z^{(1)}(\mathrm{j}, t_0\mathrm{j}; z)$ we first need $Z_1^{(0)}(1, t_0\mathrm{j}; z)$ and $Z_1^{(0)}(\mathrm{j}, t_0\mathrm{j}; z)$. These functions are calculated from the generating pair (F_1, G_1) given by

$$F_1(z) = \left(\frac{z}{2} + \mathrm{e}_1\right)\sqrt{(x+1)(t+1)},$$

$$G_1(z) = \left(\frac{\mathrm{j}z}{2} + \mathrm{e}_1\right)\frac{1}{\sqrt{(x+1)(t+1)}}.$$

Using this generating pair we obtain

$$Z_1^{(0)}(1, t_0\mathrm{j}; z) = -\frac{z + 2\mathrm{e}_1}{\alpha t_0(t_0 + 2)}\sqrt{(x+1)(t+1)}$$

$$+ \mathrm{j}\frac{\alpha(t_0 + 1)(z + 2\mathrm{e}_1)}{t_0(t_0 + 2)}\frac{1}{\sqrt{(x+1)(t+1)}},$$

$$Z_1^{(0)}(\mathrm{j}, t_0\mathrm{j}; z) = \frac{(t_0 + 1)(z + 2\mathrm{e}_1)}{\alpha t_0(t_0 + 2)}\sqrt{(x+1)(t+1)}$$

$$- \mathrm{j}\frac{\alpha(z + 2\mathrm{e}_1)}{t_0(t_0 + 2)}\frac{1}{\sqrt{(x+1)(t+1)}}.$$

We note that $F^* = -\mathrm{j}\sqrt{(x+1)(t+1)}$ and $G^* = 1/\sqrt{(x+1)(t+1)}$ such that we are now able to calculate $Z^{(1)}(1, t_0 j; z)$ and $Z^{(1)}(j, t_0 j; z)$. We find

$$
\begin{aligned}
&Z^{(1)}(1, t_0 j; z) \\
&= \sqrt{(x+1)(t+1)} \left[\frac{-1}{\alpha t_0 (t_0+2)} \left(\frac{x^2+t^2}{2} + x + t \right) + \frac{\alpha(t_0+1)}{t_0(t_0+2)} \ln[(x+1)(t+1)] \right] \\
&\quad + \frac{j}{\sqrt{(x+1)(t+1)}} \left[\frac{1}{2\alpha t_0(t_0+2)} \Big(2(x+t)(xt+1) + (x^2 t^2 + 4xt + x^2 + t^2) \Big) \right. \\
&\qquad\qquad\qquad\qquad \left. - \frac{\alpha(t_0+1)}{t_0(t_0+2)} \left(\frac{x^2+t^2}{2} + x + t \right) \right]
\end{aligned}
$$

and

$$
\begin{aligned}
&Z^{(1)}(j, t_0 j; z) \\
&= \sqrt{(x+1)(t+1)} \left[\frac{t_0+1}{\alpha t_0 (t_0+2)} \left(\frac{x^2+t^2}{2} + x + t \right) - \frac{\alpha}{t_0(t_0+2)} \ln[(x+1)(t+1)] \right] \\
&\quad + \frac{j}{\sqrt{(x+1)(t+1)}} \left[\frac{-(t_0+1)}{2\alpha t_0(t_0+2)} \Big(2(x+t)(xt+1) + (x^2 t^2 + 4xt + x^2 + t^2) \Big) \right. \\
&\qquad\qquad\qquad\qquad \left. + \frac{\alpha}{t_0(t_0+2)} \left(\frac{x^2+t^2}{2} + x + t \right) \right].
\end{aligned}
$$

Again here, we can verify that the real parts of $Z^{(1)}(1, t_0 j; z)$ and $Z^{(1)}(j, t_0 j; z)$ are solutions of the Klein-Gordon equation with potential (13.50).

Thus, we have proved that the Klein-Gordon equation can be reduced to a hyperbolic Vekua equation of the form (13.14) and as a consequence under quite general conditions an infinite system of solutions of the Klein-Gordon equation can be constructed explicitly as a real part of the corresponding set of formal powers. Meanwhile in the elliptic theory this result gave us a complete system of solutions (of a corresponding Schrödinger equation); it is an open question what part of the kernel of the Klein-Gordon operator is determined by the obtained solutions.

One of the main results of elliptic pseudoanalytic function theory is the similarity principle. It is interesting and important to find a corresponding fact in the hyperbolic case.

The reduction of the Klein-Gordon equation with an arbitrary potential to a Vekua-type hyperbolic first-order equation gives us the possibility to apply concepts and ideas from pseudoanalytic function theory to linear second-order wave equations. Besides some first applications presented here, questions related to initial and boundary value problems, existence and construction of special classes of solutions, large-time behavior of solutions (closely related to a similarity principle) and others may receive a new development effort.

Part V

Bicomplex and Biquaternionic Pseudoanalytic Functions and Applications

Chapter 14

The Dirac Equation

The Dirac equation with a fixed energy and the Vekua equation describing pseudoanalytic functions both are first-order elliptic systems, and it would be quite natural to expect a deep interrelation between their theories, especially in the case when all potentials and wave functions in the Dirac equation depend on two space variables only. Nevertheless there has not been much work done in this direction[1] due to the fact that traditional matrix representations of the Dirac operator do not allow us to visualize a relation between the Dirac equation in the two-dimensional case and the Vekua equation. Written using the traditional matrix formalism, the Dirac equation is a system of four complex equations which does not decouple in a two-dimensional situation but decouples in the one-dimensional case only.

In the present chapter we establish a simple relation between the Dirac equation with a scalar and an electromagnetic potential in a two-dimensional case from one side and a pair of decoupled Vekua equations from the other. As a first step we use the matrix transformation proposed in [65] (see also [66] and [81]) which allows us to rewrite the Dirac equation in a covariant form as a biquaternionic equation. It is not our aim to discuss here the advantages of our biquaternionic reformulation of the Dirac equation compared with other representations (the interested reader can find some of the arguments in [66]). We point out only that our transformation is $\mathbb{C}$-linear as well as is the resulting Dirac operator, which is not the case for a better known biquaternionic reformulation of the Dirac operator introduced by C. Lanczos in [83] (see [40] and [66] for more references). Moreover, in the time-dependent case, with a vanishing electromagnetic potential our Dirac operator is real quaternionic. We mention also that the quaternionic form of the Dirac equation introduced in [65] and discussed here, was recently rediscovered in [113].

Here we exploit another attractive facet of our biquaternionic Dirac equation (which, we emphasize, is completely equivalent to the traditional Dirac equation

---

[1]We refer to the work [5] where the theory of pseudoanalytic functions was used in a way completely different from ours for studying the two-dimensional Dirac equation with a scalar or a pseudoscalar potential.

written using Dirac matrices). In the two-dimensional case it decouples into a pair of separate Vekua equations. In general these Vekua equations are bicomplex. We show that some results from theory of pseudoanalytic functions without modifications can be applied to these equations. We formulate the similarity principle and concentrate on another non-trivial and surprising consequence of the established relation with pseudoanalytic functions. Consider the Dirac equation with a scalar potential depending on one variable with fixed energy and mass. In general this equation cannot be solved explicitly even if one looks for wave functions of one variable. Nonetheless one of the results we present here for such a Dirac equation is an algorithmically simple procedure for obtaining in an explicit form an infinite system of exact solutions depending on two variables. The solutions are positive formal powers and as such, in general, they are not appropriate for studying the Dirac equation on the whole plane. However the very fact that it is always possible to obtain explicitly an infinite system of exact solutions of the Dirac equation with scalar potential of one variable, as well as the hope to be able to obtain explicitly not only the generalizations of positive powers but also those of the negative ones, makes in our opinion this approach attractive and promising.

14.1   Notation

We denote by $\mathbb{H}(\mathbb{C})$ the algebra of complex quaternions (= biquaternions). The elements of $\mathbb{H}(\mathbb{C})$ have the form $Q = \sum_{k=0}^{3} Q_k e_k$ where $\{Q_k\} \subset \mathbb{C}$, e_0 is the unit and $\{e_k | k = 1, 2, 3\}$ are the standard quaternionic imaginary units. Sometimes we will use also another notation for the quaternionic imaginary units:

$$e_1 = \mathbf{i}, \quad e_2 = \mathbf{j}, \quad e_3 = \mathbf{k}.$$

We denote the imaginary unit in $\mathbb{C}$ by i as usual. By definition i commutes with e_k, $k = \overline{0,3}$. We will use also the vector representation of $Q \in \mathbb{H}(\mathbb{C})$: $Q = Sc(Q) + \mathrm{Vec}(Q)$, where $Sc(Q) = Q_0$ and $\mathrm{Vec}(Q) = \mathbf{Q} = \sum_{k=1}^{3} Q_k e_k$. The quaternionic conjugation is defined as $\overline{Q} = Q_0 - \mathbf{Q}$.

By M^P we denote the operator of multiplication by a biquaternion P from the right-hand side

$$M^P Q = Q \cdot P.$$

The interested reader can find more information on complex quaternions in, e.g., [66] or [81].

Let Q be a complex quaternion-valued differentiable function of $\mathbf{x} = (x_1, x_2, x_3)$. Let

$$DQ = \sum_{k=1}^{3} e_k \frac{\partial}{\partial x_k} Q. \tag{14.1}$$

This operator is called sometimes the Moisil-Theodorescu operator or the Dirac operator but the truth is that it was introduced already by W.R. Hamilton himself and studied in a great number of works (see, e.g., [39], [41], [42], [66], [81]). Expression (14.1) can be rewritten in a vector form as

$$DQ = -\operatorname{div}\mathbf{Q} + \operatorname{grad}Q_0 + \operatorname{rot}\mathbf{Q}.$$

That is, $\operatorname{Sc}(DQ) = -\operatorname{div}\mathbf{Q}$ and $\operatorname{Vec}(DQ) = \operatorname{grad}Q_0 + \operatorname{rot}\mathbf{Q}$. Let us notice that $D^2 = -\Delta$. If Q_0 is a scalar function, then DQ_0 coincides with $\operatorname{grad}Q_0$.

The following generalization of Leibniz's rule can be proved by a direct calculation (see [41, p. 24]).

Theorem 134. *Let $\{P, Q\} \subset C^1(G; \mathbb{H}(\mathbb{C}))$, where G is some domain in $\mathbb{R}^3$. Then*

$$D[P \cdot Q] = D[P] \cdot Q + \overline{P} \cdot D[Q] + 2(\operatorname{Sc}(PD))[Q], \qquad (14.2)$$

where

$$(\operatorname{Sc}(PD))[Q] := -\sum_{j=1}^{3} P_j \partial_j Q.$$

Remark 135. If in Theorem 134 $\operatorname{Vec}(P) = 0$, that is $P = P_0$, then

$$D[P_0 \cdot Q] = D[P_0] \cdot Q + P_0 \cdot D[Q]. \qquad (14.3)$$

From this equality we obtain that the operator $D + \frac{\operatorname{grad}P_0}{P_0}$ can be factorized as

$$\left(D + \frac{\operatorname{grad}P_0}{P_0}\right)Q = P_0^{-1}D(P_0 Q). \qquad (14.4)$$

Let $\mathbf{G}$ be a complex-valued vector such that $\operatorname{rot}\mathbf{G} \equiv 0$. Then the complex-valued scalar function φ is said to be its potential (or antigradient) if $\operatorname{grad}\varphi = \mathbf{G}$. We will write $\varphi - \mathcal{A}[\mathbf{G}]$. The operator $\mathcal{A}$ is a simple generalization of the usual antiderivative and of the operator $\overline{A}$ (see Subsection 2.3.3), and it defines the function φ up to an arbitrary constant. Its explicit representation is well known and has the form

$$\mathcal{A}[\mathbf{G}](x, y, z) = \int_{x_0}^{x} G_1(\xi, y_0, z_0)d\xi + \int_{y_0}^{y} G_2(x, \zeta, z_0)d\zeta + \int_{z_0}^{z} G_3(x, y, \eta)d\eta + C.$$

14.2   Quaternionic form of the Dirac equation

Consider the Dirac operator with scalar and electromagnetic potentials

$$\mathbb{D} = \gamma_0 \partial_t + \sum_{k=1}^{3} \gamma_k \partial_k + i\left(m + p_{el}\gamma_0 + \sum_{k=1}^{3} A_k \gamma_k + p_{sc}\right)$$

where γ_j, $j = 0, 1, 2, 3$ are usual γ-matrices (see, e.g., [16], [115])

$$\gamma_0 := \begin{pmatrix} 1 & 0 & 0 & 0 \\ 0 & 1 & 0 & 0 \\ 0 & 0 & -1 & 0 \\ 0 & 0 & 0 & -1 \end{pmatrix}, \qquad \gamma_1 := \begin{pmatrix} 0 & 0 & 0 & -1 \\ 0 & 0 & -1 & 0 \\ 0 & 1 & 0 & 0 \\ 1 & 0 & 0 & 0 \end{pmatrix},$$

$$\gamma_2 := \begin{pmatrix} 0 & 0 & 0 & i \\ 0 & 0 & -i & 0 \\ 0 & -i & 0 & 0 \\ i & 0 & 0 & 0 \end{pmatrix}, \qquad \gamma_3 := \begin{pmatrix} 0 & 0 & -1 & 0 \\ 0 & 0 & 0 & 1 \\ 1 & 0 & 0 & 0 \\ 0 & -1 & 0 & 0 \end{pmatrix},$$

$m \in \mathbb{R}$, p_{el}, A_k and p_{sc} are real-valued functions.

In [65] (see also [22], [66], [81]) a simple matrix transformation was obtained which allows us to rewrite the classical Dirac equation in quaternionic terms.

Let us introduce an auxiliary notation $\widetilde{f} := f(t, x_1, x_2, -x_3)$. The domain $\widetilde{G}$ is assumed to be obtained from the domain $G \subset \mathbb{R}^4$ by the reflection $x_3 \to -x_3$. The transformation announced above we denote as $\mathcal{K}$ and define it in the following way. A function $\Phi : G \subset \mathbb{R}^4 \to \mathbb{C}^4$ is transformed into a function $F : \widetilde{G} \subset \mathbb{R}^4 \to \mathbb{H}(\mathbb{C})$ by the rule

$$F = \mathcal{K}[\Phi] := \frac{1}{2}\left(-(\widetilde{\Phi}_1 - \widetilde{\Phi}_2)e_0 + i(\widetilde{\Phi}_0 - \widetilde{\Phi}_3)e_1 - (\widetilde{\Phi}_0 + \widetilde{\Phi}_3)e_2 + i(\widetilde{\Phi}_1 + \widetilde{\Phi}_2)e_3\right).$$

The inverse transformation $\mathcal{K}^{-1}$ is defined as

$$\Phi = \mathcal{K}^{-1}[F] = (-i\widetilde{F}_1 - \widetilde{F}_2, -\widetilde{F}_0 - i\widetilde{F}_3, \widetilde{F}_0 - i\widetilde{F}_3, i\widetilde{F}_1 - \widetilde{F}_2)^T.$$

Let us present the introduced transformations in a more explicit matrix form which relates the components of a $\mathbb{C}^4$-valued function Φ with the components of an $\mathbb{H}(\mathbb{C})$-valued function F:

$$F = \mathcal{K}[\Phi] = \frac{1}{2}\begin{pmatrix} 0 & -1 & 1 & 0 \\ i & 0 & 0 & -i \\ -1 & 0 & 0 & -1 \\ 0 & i & i & 0 \end{pmatrix}\begin{pmatrix} \widetilde{\Phi}_0 \\ \widetilde{\Phi}_1 \\ \widetilde{\Phi}_2 \\ \widetilde{\Phi}_3 \end{pmatrix}$$

and

$$\Phi = \mathcal{K}^{-1}[F] = \begin{pmatrix} 0 & -i & -1 & 0 \\ -1 & 0 & 0 & -i \\ 1 & 0 & 0 & -i \\ 0 & i & -1 & 0 \end{pmatrix}\begin{pmatrix} \widetilde{F}_0 \\ \widetilde{F}_1 \\ \widetilde{F}_2 \\ \widetilde{F}_3 \end{pmatrix}.$$

Let

$$R = D - \partial_t M^{e_1} + \mathbf{a} + M^{-i(\widetilde{p}_{el}e_1 - i(\widetilde{p}_{sc} + m)e_2)}$$

where $\mathbf{a} = i(\widetilde{A}_1 e_1 + \widetilde{A}_2 e_2 - \widetilde{A}_3 e_3)$. The following equality holds [66]:

$$R = \mathcal{K}\gamma_1\gamma_2\gamma_3 \mathbb{D}\mathcal{K}^{-1}.$$

That is, a $\mathbb{C}^4$-valued function Φ is a solution of the equation

$$\mathbb{D}\Phi = 0 \qquad \text{in } G$$

iff the complex quaternionic function $F = \mathcal{K}\Phi$ is a solution of the quaternionic equation

$$RF = 0 \qquad \text{in } \widetilde{G}.$$

Note that in the absence of the electromagnetic potential the operator R becomes real quaternionic which is an important property (see [79]).

In what follows we assume that potentials are time-independent and consider solutions with fixed energy: $\Phi(t, \mathbf{x}) = \Phi_\omega(\mathbf{x}) e^{i\omega t}$. The equation for Φ_ω has the form

$$\mathbb{D}_\omega \Phi_\omega = 0 \qquad \text{in } \widehat{G} \tag{14.5}$$

where $\widehat{G}$ is a domain in $\mathbb{R}^3$,

$$\mathbb{D}_\omega = i\omega\gamma_0 + \sum_{k=1}^{3} \gamma_k \partial_k + i\left(m + p_{el}\gamma_0 + \sum_{k=1}^{3} A_k \gamma_k + p_{sc}\right).$$

We have

$$R_\omega = \mathcal{K}\gamma_1\gamma_2\gamma_3 \mathbb{D}_\omega \mathcal{K}^{-1},$$

where

$$R_\omega = D + \mathbf{a} + M^{\mathbf{b}}$$

with $\mathbf{b} = -i((\widetilde{p}_{el} + \omega)e_1 - i(\widetilde{p}_{sc} + m)e_2)$. Thus, equation (14.5) turns into the complex quaternionic equation

$$R_\omega q = 0 \tag{14.6}$$

where q is a complex quaternion-valued function.

14.3 The Dirac equation in a two-dimensional case as a bicomplex Vekua equation

Let us introduce the following notation. For any complex quaternion q we denote by Q_1 and Q_2 its bicomplex components

$$Q_1 = q_0 + q_3 e_3 \qquad \text{and} \qquad Q_2 = q_2 - q_1 e_3.$$

Then q can be represented as $q = Q_1 + Q_2 e_2$. For the operator D we have $D = D_1 + D_2 e_2$ with $D_1 = e_3 \partial_3$ and $D_2 = \partial_2 - \partial_1 e_3$. Notice that $\mathbf{b} = Be_2$ with $B = -(\widetilde{p}_{sc} + m) + i(\widetilde{p}_{el} + \omega)e_3$, $\mathbf{a} = A_1 + A_2 e_2$ with $A_1 = a_3 e_3$ and $A_2 = a_2 - a_1 e_3$.

We obtain that equation (14.6) is equivalent to the system

$$D_1 Q_1 - D_2 \overline{Q}_2 + A_1 Q_1 - A_2 \overline{Q}_2 - \overline{B} Q_2 = 0, \tag{14.7}$$

$$D_2 \overline{Q}_1 + D_1 Q_2 + A_2 \overline{Q}_1 + A_1 Q_2 + B Q_1 = 0, \tag{14.8}$$

where Q_1 and Q_2 are bicomplex components of q. We stress that the system (14.7), (14.8) is equivalent to the Dirac equation in γ-matrices (14.5).

Let us suppose all fields in our model to be independent of x_3, and $A_1 = a_3 e_3 \equiv 0$. Then the system (14.7), (14.8) decouples, and we obtain two separate bicomplex equations [21]

$$\overline{D}_2 Q_2 = -\overline{A}_2 Q_2 - B \overline{Q}_2$$

and

$$\overline{D}_2 Q_1 = -\overline{A}_2 Q_1 - \overline{B} Q_1.$$

Write $\overline{\partial} = \overline{D}_2$, $a = -\overline{A}_2$, $b = -B$, $w = Q_2$, $W = Q_1$, $z = x + y\mathbf{k}$, where $x = x_2$, $y = x_1$ and for convenience we set $\mathbf{k} = e_3$. Then we reduce the Dirac equation with electromagnetic and scalar potentials independent of x_3 to a pair of Vekua-type equations

$$\overline{\partial} w = aw + b\overline{w} \tag{14.9}$$

and

$$\overline{\partial} W = aW + \overline{b} \overline{W}. \tag{14.10}$$

The difference between the bicomplex equations (14.9), (14.10) and the usual complex Vekua equations is revealed if only w or W can take values equal to bicomplex zero divisors (otherwise equations (14.9), (14.10) can be analyzed following Bers-Vekua theory). Let us study this possibility with the aid of the pair of projection operators

$$P^+ = \frac{1}{2}(1 + i\mathbf{k}) \qquad \text{and} \qquad P^- = \frac{1}{2}(1 - i\mathbf{k}).$$

The set of bicomplex zero divisors, that is of nonzero elements $q = q_0 + q_1 \mathbf{k}$, $\{q_0, q_1\} \subset \mathbb{C}$ such that

$$q\overline{q} = (q_0 + q_1 \mathbf{k})(q_0 - q_1 \mathbf{k}) = 0 \tag{14.11}$$

we denote by $\mathfrak{S}$.

Lemma 136. *Let q be a bicomplex number of the form $q = q_0 + q_1 \mathbf{k}$, $\{q_0, q_1\} \subset \mathbb{C}$. If $q \in \mathfrak{S}$, then $q = 2P^+ q_0$ or $q = 2P^- q_0$.*

Proof. From (14.11) it follows that $q_0^2 + q_1^2 = 0$ which gives us that $q_1 = \pm i q_0$. That is $q = q_0(1 + i\mathbf{k})$ or $q = q_0(1 - i\mathbf{k})$. $\square$

For other results on bicomplex numbers we refer to [111].

The similarity principle in general is not valid in a bicomplex situation. Nevertheless if the bicomplex pseudoanalytic function is such that its values are not zero divisors at any point of the domain of interest, then for such a function the similarity principle can be formulated without any modification (see Section 4.3).

14.4 Some definitions and results from Bers' theory for bicomplex pseudoanalytic functions

14.4.1 Generating pair, derivative and antiderivative

The definitions in the bicomplex case are quite analogous to those in the complex situation.

Definition 137. A pair of bicomplex functions $F = F_0 + F_1\mathbf{k}$ and $G = G_0 + G_1\mathbf{k}$, possessing in Ω partial derivatives with respect to the real variables x and y, is said to be a generating pair if it satisfies the inequality

$$\mathrm{Vec}(\overline{F}G) \neq 0 \qquad \text{in } \Omega.$$

The following expressions are called characteristic coefficients of the pair (F, G):

$$a_{(F,G)} = -\frac{\overline{F}G_{\overline{z}} - F_{\overline{z}}\overline{G}}{F\overline{G} - \overline{F}G}, \qquad b_{(F,G)} = \frac{FG_{\overline{z}} - F_{\overline{z}}G}{F\overline{G} - \overline{F}G},$$

$$A_{(F,G)} = -\frac{\overline{F}G_z - F_z\overline{G}}{F\overline{G} - \overline{F}G}, \qquad B_{(F,G)} = \frac{FG_z - F_zG}{F\overline{G} - \overline{F}G},$$

where the subindex $\overline{z}$ or z means the application of $\overline{\partial}$ or ∂ respectively.

Every bicomplex function W defined in a subdomain of Ω admits the unique representation $W = \phi F + \psi G$ where the functions ϕ and ψ are complex-valued.

The (F, G)-derivative $\dot{W} = \frac{d_{(F,G)}W}{dz}$ of a function W with ϕ and ψ possessing continuous partial derivatives, exists and has the form

$$\dot{W} = \phi_z F + \psi_z G = W_z - A_{(F,G)}W - B_{(F,G)}\overline{W} \qquad (14.12)$$

if and only if

$$\phi_{\overline{z}} F + \psi_{\overline{z}} G = 0. \qquad (14.13)$$

This last equation can be rewritten in the form

$$W_{\overline{z}} = a_{(F,G)}W + b_{(F,G)}\overline{W}$$

which we call the bicomplex Vekua equation. Solutions of this equation are called bicomplex (F, G)-pseudoanalytic functions.

Remark 138. The functions F and G are bicomplex (F, G)-pseudoanalytic, and $\dot{F} \equiv \dot{G} \equiv 0$.

Definition 139. Let (F, G) and (F_1, G_1) be two generating pairs in Ω. (F_1, G_1) is called the successor of (F, G) and (F, G) is called the predecessor of (F_1, G_1) if

$$a_{(F_1,G_1)} = a_{(F,G)} \qquad \text{and} \qquad b_{(F_1,G_1)} = -B_{(F,G)}.$$

By analogy with the complex case we have the following statement.

Theorem 140. *Let W be a bicomplex (F, G)-pseudoanalytic function and let (F_1, G_1) be a successor of (F, G). Then $\dot{W}$ is a bicomplex (F_1, G_1)-pseudoanalytic function.*

Definition 141. Let (F, G) be a generating pair. Its adjoint generating pair

$$(F, G)^* = (F^*, G^*)$$

is defined by the formulas

$$F^* = -\frac{2\overline{F}}{F\overline{G} - \overline{F}G}, \qquad G^* = \frac{2\overline{G}}{F\overline{G} - \overline{F}G}.$$

The (F, G)-integral is defined as

$$\int_\Gamma W d_{(F,G)} z = \frac{1}{2} \left(F(z_1) \operatorname{Sc} \int_\Gamma G^* W dz + G(z_1) \operatorname{Sc} \int_\Gamma F^* W dz \right)$$

where Γ is a rectifiable curve leading from z_0 to z_1.

If $W = \phi F + \psi G$ is a bicomplex (F, G)-pseudoanalytic function where ϕ and ψ are complex-valued functions, then

$$\int_{z_0}^{z} \dot{W} d_{(F,G)} z = W(z) - \phi(z_0)F(z) - \psi(z_0)G(z), \qquad (14.14)$$

and as $\dot{F} = \dot{G} = 0$, this integral is path-independent and represents the (F, G)-antiderivative of $\dot{W}$.

14.4.2 Generating sequences and Taylor series in formal powers

Definition 142. A sequence of generating pairs $\{(F_m, G_m)\}$, $m = 0, \pm 1, \pm 2, \ldots$, is called a generating sequence if (F_{m+1}, G_{m+1}) is a successor of (F_m, G_m). If $(F_0, G_0) = (F, G)$, we say that (F, G) is embedded in $\{(F_m, G_m)\}$.

Theorem 143. *Let (F, G) be a generating pair in Ω. Let Ω_1 be a bounded domain, $\overline{\Omega}_1 \subset \Omega$. Then (F, G) can be embedded in a generating sequence in Ω_1.*

Definition 144. A generating sequence $\{(F_m, G_m)\}$ is said to have period $\mu > 0$ if $(F_{m+\mu}, G_{m+\mu})$ is equivalent to (F_m, G_m), that is their characteristic coefficients coincide.

Let W be a bicomplex (F, G)-pseudoanalytic function. Using a generating sequence in which (F, G) is embedded we can define the higher derivatives of W by the recursion formula

$$W^{[0]} = W; \qquad W^{[m+1]} = \frac{d_{(F_m, G_m)} W^{[m]}}{dz}, \qquad m = 1, 2, \ldots.$$

Definition 145. The formal power $Z_m^{(0)}(a, z_0; z)$ with center at $z_0 \in \Omega$, coefficient a and exponent 0 is defined as the linear combination of the generators F_m, G_m with complex constant coefficients λ, μ chosen so that $\lambda F_m(z_0) + \mu G_m(z_0) = a$. The formal powers with exponents $n = 1, 2, \ldots$ are defined by the recursion formula

$$Z_m^{(n+1)}(a, z_0; z) = (n+1) \int_{z_0}^{z} Z_{m+1}^{(n)}(a, z_0; \zeta) d_{(F_m, G_m)} \zeta. \tag{14.15}$$

This definition implies the following properties.

1. $Z_m^{(n)}(a, z_0; z)$ is an (F_m, G_m)-pseudoanalytic function of z.

2. If a' and a'' are complex constants, then

$$Z_m^{(n)}(a' + \mathbf{k}a'', z_0; z) = a' Z_m^{(n)}(1, z_0; z) + a'' Z_m^{(n)}(\mathbf{k}, z_0; z).$$

3. The formal powers satisfy the differential relations

$$\frac{d_{(F_m, G_m)} Z_m^{(n)}(a, z_0; z)}{dz} = n Z_{m+1}^{(n-1)}(a, z_0; z).$$

4. The asymptotic formulas

$$Z_m^{(n)}(a, z_0; z) \sim a(z - z_0)^n, \quad z \to z_0$$

hold.

Assume now that

$$W(z) = \sum_{n=0}^{\infty} Z^{(n)}(a, z_0; z) \tag{14.16}$$

where the absence of the subindex m means that all the formal powers correspond to the same generating pair (F, G), and the series converges uniformly in some neighborhood of z_0. If the function W in (14.16) is bicomplex (F, G)-pseudoanalytic and the series converges normally, the rth derivative of W admits the expansion

$$W^{[r]}(z) = \sum_{n=r}^{\infty} n(n-1) \cdots (n-r+1) Z_r^{(n-r)}(a_n, z_0; z).$$

From this the Taylor formulas for the coefficients are obtained

$$a_n = \frac{W^{[n]}(z_0)}{n!}. \tag{14.17}$$

Definition 146. Let $W(z)$ be a given (F, G)-pseudoanalytic function defined for small values of $|z - z_0|$. The series

$$\sum_{n=0}^{\infty} Z^{(n)}(a, z_0; z) \tag{14.18}$$

with the coefficients given by (4.13) is called the Taylor series of W at z_0, formed with formal powers.

The Taylor series always represents the function asymptotically:

$$W(z) - \sum_{n=0}^{N} Z^{(n)}(a, z_0; z) = O\left(|z - z_0|^{N+1}\right), \quad z \to z_0, \tag{14.19}$$

for all N. This implies (since a pseudoanalytic function can not have a zero of arbitrarily high order without vanishing identically) that the sequence of derivatives $\left\{W^{[n]}(z_0)\right\}$ determines the function W uniquely.

In spite of a complete structural similarity of the complex and bicomplex main pseudoanalytic definitions, the theorems establishing the numerous properties of series in formal powers remain unproved in the case of bicomplex pseudoanalytic function theory due to the above-mentioned difficulties with the similarity principle.

14.5   The main bicomplex Vekua equation

As in the complex case we consider the main bicomplex Vekua equation. Let f_0 be a complex-valued (with respect to i), twice differentiable nonvanishing function defined on Ω. Consider the equation

$$\overline{\partial} W = \frac{\overline{\partial} f_0}{f_0} \overline{W} \quad \text{in } \Omega \tag{14.20}$$

where we recall that $\overline{\partial} = \partial_x + \mathbf{k}\partial_y$ and $W = W_1 + \mathbf{k}W_2$, $W_{1,2}$ are complex-valued (with respect to i).

Set $\nu_1 = \Delta f_0 / f_0$.

Theorem 147 ([71]). *If $W = W_1 + W_2\mathbf{k}$ is a solution of* (14.20)*, then $W_1 = \mathrm{Sc}\, W$ is a solution of the stationary Schrödinger equation*

$$-\Delta W_1 + \nu_1 W_1 = 0 \quad \text{in } \Omega \tag{14.21}$$

and $W_2 = \mathrm{Vec}\, W$ is a solution of the associated Schrödinger equation

$$-\Delta W_2 + \nu_2 W_2 = 0 \quad \text{in } \Omega \tag{14.22}$$

where $\nu_2 = 2(\overline{\partial} f_0 \cdot \partial f_0)/f_0^2 - \nu_1$.

Moreover, the results from Section 3.3 allowing us by a given solution W_1 of (14.21) to construct such a solution W_2 of (14.22) that $W = W_1 + W_2\mathbf{k}$ be a solution of (14.20) generalizing in this way the well-known procedure for constructing conjugate harmonic functions in complex analysis remain valid in the studied bicomplex case as well.

14.6 Dirac equation with a scalar potential

Let us show that the Dirac equation with a scalar potential depending on one real variable reduces to a bicomplex Vekua equation of the form (14.20). In this section we present results from [21].

Let $p_{sc} = p(x)$ and $p_{el} \equiv 0$, $A_k \equiv 0$, $k = 1, 2, 3$. Then according to Section 14.3 the Dirac equation is equivalent to the pair of bicomplex Vekua equations

$$\overline{\partial} w = b\overline{w} \tag{14.23}$$

and

$$\overline{\partial} W = \overline{b}\overline{W} \tag{14.24}$$

with $b = p(x) + m - i\omega\mathbf{k}$.

Let $f_0 = e^{P(x)+mx+i\omega y}$, where P is an antiderivative of p. Then we have

$$\overline{b} = \overline{\partial} f_0 / f_0.$$

Note that due to Theorem 147, if the bicomplex function W is a solution of (14.24), then the complex function $W_1 = \mathrm{Sc}\, W$ is a solution of the stationary Schrödinger equation (14.21) where

$$\nu_1(x) = p'(x) + (p(x) + m)^2 - \omega^2, \tag{14.25}$$

and the function $W_2 = \mathrm{Vec}\, W$ is a solution of equation (14.22) where

$$\nu_2(x) = -p'(x) + (p(x) + m)^2 - \omega^2. \tag{14.26}$$

Let us notice that both Schrödinger equations (14.21) and (14.22) in this case admit separation of variables. Nevertheless this does not imply they can be solved explicitly. In general this is not the case. However we will show how using our approach and Bers' theory for both of them one can construct in explicit form an infinite system of exact solutions.

Consider equation (14.24). It is easy to see that the pair of functions

$$F = f_0 \quad \text{and} \quad G = \frac{\mathbf{k}}{f_0} \tag{14.27}$$

represents a generating pair for (14.24). Note that $F = e^\sigma$ and $G = e^{-\sigma}\mathbf{k}$, where $\sigma = \alpha(x) + \beta(y)$ and $\alpha(x) = P(x) + mx$, $\beta(y) = i\omega y$. For a generating pair of such special kind it is easy to construct a successor. Let $\tau = -\alpha(x) + \beta(y)$. Then the pair $F_1 = e^\tau$ and $G_1 = e^{-\tau}\mathbf{k}$ is a successor of (F, G). Moreover, (F, G) is a successor of (F_1, G_1). Thus, for (F, G) we obtain a complete periodic generating sequence of a period 2 in explicit form.

The fact that we have a generating sequence in an explicit form implies that we are able to construct the corresponding formal powers of any order explicitly and therefore to obtain an infinite system of exact solutions of the Dirac equation with a scalar potential depending on one variable as well as of the stationary Schrödinger equations (14.21) and (14.22) with potentials (14.25) and (14.26) respectively.

As a first step we construct the adjoint generating pair (see Definition 141):

$$F^* = -f_0\mathbf{k} \qquad \text{and} \qquad G^* = \frac{1}{f_0}.$$

Next, we write down the expression for the (F, G)-integral:

$$\int_\Gamma W d_{(F,G)}z = \frac{1}{2}\left(f_0(z_1)\,\text{Sc}\int_\Gamma \frac{W(z)}{f_0(z)}dz - \frac{\mathbf{k}}{f_0(z_1)}\,\text{Sc}\int_\Gamma f_0(z)W(z)\mathbf{k}dz \right).$$

By definition, the formal power $Z^{(0)}(a, z_0; z)$ for equation (14.24) has the form

$$Z^{(0)}(a, z_0; z) = \lambda F(z) + \mu G(z),$$

where the complex constants λ and μ are chosen so that $\lambda F(z_0) + \mu G(z_0) = a$. That is,

$$Z^{(0)}(a, z_0; z) = \lambda e^{P(x)+mx+i\omega y} + \mu e^{-(P(x)+mx+i\omega y)}\mathbf{k}.$$

In order to obtain $Z^{(1)}(a, z_0; z)$ we should take the (F, G)-integral of $Z_1^{(0)}(a, z_0; z)$, where

$$Z_1^{(0)}(a, z_0; z) = \lambda_1 F_1(z) + \mu_1 G_1(z),$$

with $\lambda_1 F_1(z_0) + \mu_1 G_1(z_0) = a$. Thus,

$$Z^{(1)}(a, z_0; z) = \int_{z_0}^z (\lambda_1 F_1(\zeta) + \mu_1 G_1(\zeta))d_{(F,G)}\zeta$$

$$= \frac{1}{2}\bigg\{ e^{P(x)+mx+i\omega y}\,\text{Sc}$$

$$\times \int_{z_0}^z e^{-P(x')-mx'-i\omega y'}(\lambda_1 e^{-P(x')-mx'+i\omega y'} + \mu_1 e^{P(x')+mx'-i\omega y'}\mathbf{k})d\zeta$$

$$- e^{-P(x)-mx-i\omega y}\mathbf{k}\,\text{Sc}$$

$$\times \int_{z_0}^z e^{P(x')+mx'+i\omega y'}\mathbf{k}(\lambda_1 e^{-P(x')-mx'+i\omega y'} + \mu_1 e^{P(x')+mx'-i\omega y'}\mathbf{k})d\zeta \bigg\}$$

$$= \frac{1}{2}\left\{ e^{P(x)+mx+i\omega y}\operatorname{Sc}\int_{z_0}^{z}(\lambda_1 e^{-2(P(x')+mx')} + \mu_1 e^{-2i\omega y'}\mathbf{k})d\zeta \right.$$

$$\left. - e^{-P(x)-mx-i\omega y}\mathbf{k}\operatorname{Sc}\int_{z_0}^{z}(\lambda_1 e^{2i\omega y'}\mathbf{k} - \mu_1 e^{2(P(x')+mx')})d\zeta \right\}$$

where $\zeta = x' + y'\mathbf{k}$.

For $Z^{(2)}(a, z_0; z)$ by Definition 145 we have

$$Z^{(2)}(a, z_0; z) = 2\int_{z_0}^{z} Z_1^{(1)}(a, z_0; \zeta)d_{(F,G)}\zeta, \qquad (14.28)$$

where $Z_1^{(1)}(a, z_0; \zeta)$ in its turn can be found from the equality

$$Z_1^{(1)}(a, z_0; z) = \int_{z_0}^{z} Z_2^{(0)}(a, z_0; \zeta)d_{(F_1,G_1)}\zeta. \qquad (14.29)$$

We note that due to periodicity of the generating sequence containing the generating pair (14.27),

$$Z_2^{(0)}(a, z_0; \zeta) = Z^{(0)}(a, z_0; \zeta).$$

The adjoint pair for (F_1, G_1) necessary for the (F_1, G_1)-integral in (14.29) has the form

$$F_1^* = -e^{\tau}\mathbf{k} \qquad \text{and} \qquad G_1^* = e^{-\tau}.$$

Thus,

$$Z_1^{(1)}(a, z_0; z)$$

$$= \frac{1}{2}\left\{ e^{-P(x)-mx+i\omega y}\operatorname{Sc} \right.$$

$$\times \int_{z_0}^{z} e^{P(x')+mx'-i\omega y'}(\lambda e^{P(x')+mx'+i\omega y'} + \mu e^{-P(x')-mx'-i\omega y'}\mathbf{k})d\zeta$$

$$- e^{P(x)+mx-i\omega y}\mathbf{k}\operatorname{Sc}$$

$$\left. \times \int_{z_0}^{z} e^{-P(x')-mx'+i\omega y'}\mathbf{k}(\lambda e^{P(x')+mx'+i\omega y'} + \mu e^{-P(x')-mx'-i\omega y'}\mathbf{k})d\zeta \right\}$$

$$= \frac{1}{2}\left\{ e^{-P(x)-mx+i\omega y}\operatorname{Sc}\int_{z_0}^{z}(\lambda e^{2(P(x')+mx')} + \mu e^{-2i\omega y'}\mathbf{k})d\zeta \right.$$

$$\left. - e^{P(x)+mx-i\omega y}\mathbf{k}\operatorname{Sc}\int_{z_0}^{z}(\lambda e^{2i\omega y'}\mathbf{k} - \mu e^{-2(P(x')+mx')})d\zeta \right\}.$$

Substitution of this expression into (14.28) gives us the formal power $Z^{(2)}(a, z_0; z)$, and this algorithmically simple procedure can be continued indefinitely. As a result we obtain an infinite system of formal powers of (14.24).

A similar procedure works also for equation (14.23). Note that the pair of functions $F_1\mathbf{k} = e^\tau\mathbf{k}$ and $G_1\mathbf{k} = -e^{-\tau}$ is a generating pair corresponding to (14.23).

As any solution of the Schrödinger equation (14.21) with the potential ν_1 defined by (14.25) is the scalar part of some solution of (14.24), and any solution of (14.22) with the potential (14.26) is the vector part of some solution of (14.24), the scalar and the vector parts of the constructed system of formal powers give us infinite systems of solutions of (14.21) and (14.22) respectively.

This last result can also be interpreted in the following way. Consider the equation

$$-\Delta f + \nu f = \omega^2 f \qquad \text{in } \Omega \tag{14.30}$$

where f is a complex twice-continuously differentiable function of two real variables x and y, and ν is a complex-valued function of one real variable x, ω is a complex constant. Suppose we are given a particular solution $f_0 = f_0(x)$ of the ordinary differential equation

$$-\frac{d^2 f_0}{dx^2} + \nu f_0 = 0. \tag{14.31}$$

This implies that we are able to represent ν in the form $\nu = p' + p^2$ where $p = f_0'/f_0$. Then we observe that (14.30) is precisely equation (14.21) with $m = 0$ in (14.25). Thus our result means that, if we are able to solve the ordinary differential equation (14.31), then we can construct explicitly an infinite system of exact solutions to (14.30) for any ω. For this one should consider the bicomplex Vekua equation (14.24) and follow the procedure described above for constructing the corresponding system of formal powers. Then the scalar part of the system gives us an infinite system of exact solutions to (14.30).

Chapter 15

Complex Second-order Elliptic Equations and Bicomplex Pseudoanalytic Functions

Using the formalism of bicomplex numbers many results from Chapters 3 and 4 can be obtained also for second-order elliptic equations with complex coefficients. Here we give some examples without proofs which are completely analogous to those given earlier.

Consider the equation

$$(-\Delta + \nu) f = 0 \tag{15.1}$$

in some domain $\Omega \subset \mathbf{R}^2$, where $\Delta = \frac{\partial^2}{\partial x^2} + \frac{\partial^2}{\partial y^2}$, ν and f are complex-valued with respect to the imaginary unit i (in our terms scalar) functions. We assume that f is a twice-continuously differentiable function. We write $\partial_{\bar{z}} = \frac{1}{2}\bar{\partial} = \frac{1}{2}(\partial_x + \mathbf{k}\partial_y)$, $\partial_z = \frac{1}{2}\partial = \frac{1}{2}(\partial_x - \mathbf{k}\partial_y)$, and by C the operator of conjugation with respect to $\mathbf{k}$: if $W = W_1 + \mathbf{k}W_2$ where W_1 and W_2 are scalar functions, then $CW = W_1 - \mathbf{k}W_2$.

We have the following results on the factorization of second-order operators generalizing those of Chapter 3.

Theorem 148. *Let f be a nonvanishing in Ω particular solution of (15.1). Then for any complex-valued (scalar) function $\varphi \in C^2(\Omega)$ the following equalities hold:*

$$\frac{1}{4}(\Delta - \nu)\varphi = \left(\partial_{\bar{z}} + \frac{f_z}{f}C\right)\left(\partial_z - \frac{f_z}{f}C\right)\varphi = \left(\partial_z + \frac{f_{\bar{z}}}{f}C\right)\left(\partial_{\bar{z}} - \frac{f_{\bar{z}}}{f}C\right)\varphi. \tag{15.2}$$

Theorem 149. *Let u_0 be a nonvanishing in Ω particular solution of the equation*

$$(\operatorname{div} p \operatorname{grad} + q)u = 0 \qquad in \ \Omega \tag{15.3}$$

where p and q are complex-valued functions, $p \in C^2(\Omega)$ and $p \neq 0$ in Ω. Then for any complex-valued (scalar) twice-continuously differentiable function φ the *following equality holds:*

$$\frac{1}{4}(\operatorname{div} p \operatorname{grad} + q)\varphi = p^{1/2}\left(\partial_z + \frac{f_{\bar{z}}}{f}C\right)\left(\partial_{\bar{z}} - \frac{f_{\bar{z}}}{f}C\right)p^{1/2}\varphi,$$

where $f = p^{1/2}u_0$.

Let f be a scalar function of x and y. Consider the main bicomplex Vekua equation

$$W_{\bar{z}} = \frac{f_{\bar{z}}}{f}\overline{W} \qquad in\ \Omega. \tag{15.4}$$

Set $W_1 = \operatorname{Sc} W$ and $W_2 = \operatorname{Vec} W$.

Theorem 150. *Let* $W = W_1 + W_2\mathbf{k}$ *be a solution of* (15.4). *Then* $U = f^{-1}W_1$ *is a solution of the equation*

$$\operatorname{div}(f^2\nabla U) = 0 \qquad in\ \Omega, \tag{15.5}$$

and $V = fW_2$ *is a solution of the equation*

$$\operatorname{div}(f^{-2}\nabla V) = 0 \qquad in\ \Omega, \tag{15.6}$$

the function W_1 *is a solution of the stationary Schrödinger equation*

$$-\Delta W_1 + r_1 W_1 = 0 \qquad in\ \Omega \tag{15.7}$$

with $r_1 = \Delta f/f$, *and* W_2 *is a solution of the associated Schrödinger equation*

$$-\Delta W_2 + r_2 W_2 = 0 \qquad in\ \Omega \tag{15.8}$$

where $r_2 = 2(\nabla f)^2/f^2 - r_1$ *and* $(\nabla f)^2 = f_x^2 + f_y^2$.

Remark 151. The pair of functions

$$F = f \quad and \quad G = \frac{\mathbf{k}}{f} \tag{15.9}$$

is a generating pair for (15.4). This allows us to rewrite (15.4) in the form of an equation for pseudoanalytic functions of second kind,

$$\varphi_{\bar{z}}f + \psi_{\bar{z}}\frac{\mathbf{k}}{f} = 0, \tag{15.10}$$

where φ and ψ are scalar functions. If φ and ψ satisfy (15.10), then $W = \varphi f + \psi\frac{\mathbf{k}}{f}$ is a solution of (15.4) and vice versa.

Set $w = \varphi + \psi\mathbf{k}$. Then from (15.10) we have

$$(w + \overline{w})_{\bar{z}}f + (w - \overline{w})_{\bar{z}}\frac{1}{f} = 0,$$

which is equivalent to the equation

$$w_{\bar{z}} = \frac{1 - f^2}{1 + f^2}\overline{w}_{\bar{z}}.$$

Theorem 152. *Let $W = W_1 + W_2\mathbf{k}$ be a solution of* (15.4). *Assume that $f = p^{1/2}u_0$, where u_0 is a nonvanishing solution of* (15.3) *in Ω. Then $u = p^{-1/2}W_1$ is a solution of* (15.3) *in Ω, and $v = p^{1/2}W_2$ is a solution of the equation*

$$\left(\operatorname{div} \frac{1}{p} \operatorname{grad} + q_1 \right) v = 0 \qquad in\ \Omega, \tag{15.11}$$

where

$$q_1 = -\frac{1}{p} \left(\frac{q}{p} + 2 \left\langle \frac{\nabla p}{p}, \frac{\nabla u_0}{u_0} \right\rangle + 2 \left(\frac{\nabla u_0}{u_0} \right)^2 \right). \tag{15.12}$$

Theorem 153. *Let W_1 be a solution of* (15.7) *in a simply connected domain Ω. Then the function W_2, solution of* (15.8) *such that $W = W_1 + W_2\mathbf{k}$ is a solution of* (15.4), *is constructed according to the formula*

$$W_2 = f^{-1}\overline{A}(\mathbf{k}f^2 \partial_{\overline{z}}(f^{-1}W_1)).$$

Given a solution W_2 of (15.8), *the corresponding solution W_1 of* (15.7) *such that $W = W_1 + W_2\mathbf{k}$ is a solution of* (15.4), *is constructed as*

$$W_1 = -f\overline{A}(\mathbf{k}f^{-2} \partial_{\overline{z}}(fW_2)).$$

As we see the results concerning the relationship with second-order equations for bicomplex pseudoanalytic functions are similar in their structure to the corresponding results for complex pseudoanalytic functions. Moreover, the scheme for constructing formal powers and corresponding infinite systems of solutions of the second-order equations works without any modification in this bicomplex formalism. The difficulty arises when one wants to obtain results on the completeness of the systems of solutions. For this it is necessary to generalize many theorems from pseudoanalytic function theory onto the bicomplex situation. It should be mentioned that this task is anything but easy.

We notice also that while we have considered a bicomplex-valued function of a complex variable, the theory of pseudoanalytic bicomplex functions of a bicomplex variable represents interest as well; for some first results in this direction we refer to [109].

Chapter 16

Multidimensional Second-order Equations

16.1 Factorization

Consider the equation

$$(-\Delta + \nu)\, g = 0 \qquad \text{in } G \tag{16.1}$$

where $\Delta = \frac{\partial^2}{\partial x^2} + \frac{\partial^2}{\partial y^2} + \frac{\partial^2}{\partial z^2}$, ν and g are complex-valued functions, and G is a domain in $\mathbb{R}^3$. We assume that g is twice-continuously differentiable.

Theorem 154. *Let f be a nonvanishing particular solution of* (16.1). *Then for any scalar twice-continuously differentiable function g the following equality holds:*

$$(D + M^{\frac{Df}{f}})(D - M^{\frac{Df}{f}})g = (-\Delta + \nu)\, g. \tag{16.2}$$

Proof. This is a direct calculation based on the Leibniz rule (14.3). □

Remark 155. The factorization (16.2) was obtained in [9], [11] in a form which required a solution of an associated biquaternionic Riccati equation. In [64] it was shown that the solution has necessarily the form Df/f with f being a solution of (16.1).

Remark 156. Theorem 154 generalizes Theorem 25. In a two-dimensional situation (16.2) reduces to (3.2).

Remark 157. As g in (16.2) is a scalar function, the factorization of the Schrödinger operator can also be written in the form

$$(D + M^{\frac{Df}{f}})f D(f^{-1}g) = (-\Delta + \nu)\, g,$$

from which it is obvious that, if g is a solution of (16.1), then the vector $\mathbf{F} = fD(f^{-1}g)$ is a solution of the equation

$$(D + M^{\frac{Df}{f}})\mathbf{F} = 0 \qquad \text{in } G. \tag{16.3}$$

The inverse result we formulate as the following statement.

Theorem 158. *Let* $\mathbf{F}$ *be a solution of* (16.3) *in a simply connected domain* G. *Then* $g = f\mathcal{A}[f^{-1}\mathbf{F}]$ *is a solution of* (16.1).

Proof. First, in order to apply the operator $\mathcal{A}$ to the vector $f^{-1}\mathbf{F}$ we should ascertain that indeed,

$$\operatorname{rot}(f^{-1}\mathbf{F}) = 0. \tag{16.4}$$

For this, consider the vector part of (16.3). It has the form

$$\operatorname{rot}\mathbf{F} + [\mathbf{F} \times \frac{Df}{f}] = 0$$

which is equivalent to equation (16.4).

Now, applying the Laplacian to $g = f\mathcal{A}[f^{-1}\mathbf{F}]$ and taking into account that f is a solution of (16.1) and $\mathbf{F}$ is a solution of (16.3), we obtain the result:

$$\begin{aligned}
-\Delta g = D^2 g &= D(Df \cdot \mathcal{A}[f^{-1}\mathbf{F}] + \mathbf{F}) \\
&= f^{-1}\mathbf{F}Df - \mathcal{A}[f^{-1}\mathbf{F}]\Delta f + D\mathbf{F} \\
&= \mathbf{F}\frac{Df}{f} - \nu f\mathcal{A}[f^{-1}\mathbf{F}] - \mathbf{F}\frac{Df}{f} \\
&= -\nu g. \qquad \qquad \qquad \qquad \qquad \square
\end{aligned}$$

In the same way as in Section 3.2 we obtain the factorization of the operator $\operatorname{div} p \operatorname{grad} + q$ where div and grad are already operators with respect to three independent variables.

Theorem 159. *Let* u_0 *be a nonvanishing particular solution of the equation*

$$(\operatorname{div} p \operatorname{grad} + q)u = 0 \qquad in\ G \subset \mathbb{R}^3 \tag{16.5}$$

with p, q *and* u *being complex-valued functions,* $p \in C^2(G)$ *and* $p \neq 0$ *in* G. *Then for any scalar function* $\varphi \in C^2(G)$ *the following equality holds:*

$$(\operatorname{div} p \operatorname{grad} + q)\varphi = -p^{1/2}(D + M^{\frac{Df}{f}})(D - M^{\frac{Df}{f}})p^{1/2}\varphi \tag{16.6}$$

where $f = p^{1/2}u_0$.

Proof. This is analogous to the proof of Theorem 29. $\qquad \qquad \qquad \qquad \square$

Thus, if u is a solution of equation (16.5) then

$$\mathbf{F} = fD(f^{-1}p^{1/2}u) = fD(u_0^{-1}u)$$

is a solution of equation (16.3) (see Remark 157). The inverse result has the following form.

Theorem 160. *Let* **F** *be a solution of equation* (16.3) *in a simply connected domain* G, *where* $f = p^{1/2}u_0$ *and* u_0 *is a nonvanishing particular solution of* (16.5). *Then*

$$u = u_0 \mathcal{A}[f^{-1}\mathbf{F}]$$

is a solution of (16.5).

Proof. This is a corollary of Theorem 158 and relation $(\operatorname{div} p \operatorname{grad} + q) = p^{1/2}(\Delta - \nu)p^{1/2}$ where $\nu = \Delta f/f$. $\qquad\square$

Notice that due to the fact that in (16.6) φ is scalar, we can rewrite the equality in the form

$$(\operatorname{div} p \operatorname{grad} + q)\varphi = -p^{1/2}\left(D + M^{\frac{Df}{f}}\right)\left(D - \frac{Df}{f}C_H\right)p^{1/2}\varphi,$$

where C_H is the operator of quaternionic conjugation: $C_H W = \operatorname{Sc}(W) - \operatorname{Vec}(W)$.

16.2 The main quaternionic Vekua equation

Consider the equation

$$\left(D - \frac{Df}{f}C_H\right)W = 0, \tag{16.7}$$

where W is an $\mathbb{H}(\mathbb{C})$-valued function. Equation (16.7) is a direct generalization of the main Vekua equation (3.15). Moreover, we show that it preserves some important properties of (3.15).

Theorem 161. *Let* $W = W_0 + \mathbf{W}$ *be a solution of* (16.7). *Then* W_0 *is a solution of the stationary Schrödinger equation*

$$-\Delta W_0 + \nu W_0 = 0, \tag{16.8}$$

where $\nu = \Delta f/f$; *the function* $u = f^{-1}W_0$ *is a solution of the equation*

$$\operatorname{div}(f^2 \operatorname{grad} u) = 0, \tag{16.9}$$

and the vector function $\mathbf{v} = f\mathbf{W}$ *is a solution of the equation*

$$\operatorname{rot}(f^{-2} \operatorname{rot} \mathbf{v}) = 0. \tag{16.10}$$

Proof. Equation (16.7) is equivalent to the system

$$\operatorname{div} \mathbf{W} + \left\langle \frac{\nabla f}{f}, \mathbf{W} \right\rangle = 0,$$

$$\operatorname{rot} \mathbf{W} + \left[\frac{\nabla f}{f} \times \mathbf{W}\right] + \nabla W_0 - \frac{\nabla f}{f}W_0 = 0$$

which can be rewritten in the form

$$\operatorname{div}(f\mathbf{W}) = 0, \tag{16.11}$$

$$f^{-1}\operatorname{rot}(f\mathbf{W}) + f\operatorname{grad}(f^{-1}W_0) = 0. \tag{16.12}$$

From (16.12) we obtain (16.9) and (16.10). Equation (16.8) is obtained from (16.9) and (3.9). $\cdot$ □

Remark 162. Observe that the functions

$$F_0 = f, \quad F_1 = \frac{\mathbf{i}}{f}, \quad F_2 = \frac{\mathbf{j}}{f}, \quad F_3 = \frac{\mathbf{k}}{f}$$

give us a generating quartet for the equation (16.7): they are solutions of (16.7) and obviously any $\mathbb{H}(\mathbb{C})$-valued function W can be represented in the form

$$W = \sum_{j=0}^{3} \varphi_j F_j,$$

where φ_j are complex-valued functions. It is easy to verify that the function W is a solution of (16.7) iff

$$\sum_{j=0}^{3} (D\varphi_j) F_j = 0 \tag{16.13}$$

in a complete analogy with the two-dimensional case (see Remark 34). Set

$$w = \varphi_0 + \varphi_1 \mathbf{i} + \varphi_2 \mathbf{j} + \varphi_3 \mathbf{k}.$$

Then (16.13) can be written as

$$D(w + \overline{w})f + D(w - \overline{w})\frac{1}{f} = 0$$

which is equivalent to the equation

$$Dw = \frac{1 - f^2}{1 + f^2} D\overline{w}.$$

Remark 163. The results of this section remain valid in the n-dimensional situation if instead of quaternions the Clifford algebra $Cl_{0,n}$ (see, e.g., [19], [42]) is considered. The operator D is then introduced as $D = \sum_{j=1}^{n} e_j \frac{\partial}{\partial x_j}$ where e_j are the basis elements of the Clifford algebra.

Let us notice that some results in the direction of construction of a multidimensional theory of pseudoanalytic functions were presented, e.g., in [8], [88]. The difference of our approach consists in the fact that we start from the factorization of the stationary Schrödinger operator and study the quaternionic Vekua equation arising from this factorization. We expect that, due to a special form of this Vekua equation, more results from L. Bers' pseudoanalytic function theory can be generalized offering interesting applications to second-order equations of mathematical physics.

Open Problems

New mathematical results unfailingly produce new questions, and the theory presented in this book is not an exception. Here we summarize some of the open problems related to the material discussed in the preceding chapters.

1. In Chapter 5 the pseudoanalytic Cauchy kernel was obtained only for a certain class of Vekua equations and related systems describing p-analytic functions. Besides the Cauchy integral formula the explicit construction of a Cauchy kernel allows us to obtain negative formal powers, to develop Laurent series theory and to consider problems, for example, in unbounded domains. Though, as was shown in Chapter 4 it is possible to construct positive formal powers in a quite general situation, it is an important open problem, how under the same conditions to construct negative formal powers. It is worth mentioning that the proof of their existence given by L. Bers in [14] is constructive and reduces the problem to construction of positive formal powers. Nevertheless the procedure in Chapter 4 does not seem to be directly applicable in this case.

2. In Part IV the theory of hyperbolic pseudoanalytic functions and their relation to the Klein-Gordon equation were presented. In spite of a certain structural similarity of this theory to the elliptic pseudoanalytic function theory, many important questions remain unanswered. For example, as was shown in Section 13.3, formal powers for the main hyperbolic Vekua equation and consequently an infinite system of exact solutions for the corresponding Klein-Gordon equation can be constructed explicitly in a quite general situation. Nevertheless the completeness of these systems of functions in the kernels of the corresponding operators is an open question, and in case of the incompleteness their physical meaning is an interesting issue.

3. Part V shows the importance of bicomplex pseudoanalytic functions for studying second-order partial differential equations with complex coefficients and such physical systems as the Dirac equation. However, a good part of bicomplex pseudoanalytic function theory remains undeveloped. The global completeness of the system of formal powers (at least for the main Vekua equation) is one of the many open questions here. Under what conditions can such important results as the Liouville theorem be true? This is a kind

of problem needed to be solved for developing the mathematical theory of electrical impedance tomography [3] in the case of physical media with complex electrical parameters.

4. Many concepts and results of pseudoanalytic function theory wait for their appropriate generalization onto the multidimensional case using such tools as quaternions and Clifford algebras. While the main quaternionic Vekua equation was introduced in Section 16.2 and its relation to second-order Schrödinger-like equations was established, further questions as, for example, the construction of quaternionic formal powers remain open.

5. One of the main objects studied in this book is the main Vekua equation deeply related to second-order Schrödinger-like equations. Almost everywhere in the preceding chapters we required that the coefficient in this equation be nonsingular, that is the particular solution of the corresponding second-order equation whose logarithmic derivative represents that coefficient must have no zeros. Nevertheless it is very interesting to use the results on Vekua equations with singular coefficients like those from [92], [119] and references therein in order to study second-order equations which do not possess non-vanishing solutions in a domain of interest.

Bibliography

[1] Agmon, S. and Bers, L.; 1952. *The expansion theorem for pseudo-analytic functions.* Proc. Amer. Math. Soc. **3**, 757–764.

[2] Aleksandrov, A.Ya. and Solovyev, Yu.I.; 1978. Spatial problems of elasticity theory: application of methods of the theory of functions of a complex variable. Moscow: Nauka (in Russian).

[3] Astala, K. and Päivärinta, L.; 2006. *Calderón's inverse conductivity problem in the plane.* Annals of Mathematics, **163**, No. 1, 265–299.

[4] Athanasiadis, C., Costakis, G. and Stratis, I.G.; 2000. *On some properties of Beltrami fields in chiral media.* Reports on Mathematical Physics **45**, 257–271.

[5] Battle, G.A.; 1981. *Generalized analytic functions and the two-dimensional Euclidean Dirac operator.* Comm. in Partial Differential Equations **6** (2), 121–151.

[6] Begehr, H.; 1985. Boundary value problems for analytic and generalized analytic functions. Oxford: North Oxford Academic, "Complex analysis: methods, trends, and applications", Ed. by E. Lanckau and W. Tutschke, 150–165.

[7] Ben Amara, J. and Shkalikov, A.A.; 1999. *A Sturm-Liouville problem with physical and spectral parameters in boundary conditions.* Mathematical Notes 66, no. 2, 127–134.

[8] Berglez, P.; 2007. *On generalized derivatives and formal powers for pseudo-analytic functions.* Matematiche 62, no. 2, 29–36.

[9] Bernstein, S.; 1996. Factorization of solutions of the Schrödinger equation. In: Proceedings of the symposium *Analytical and numerical methods in quaternionic and Clifford analysis*, Seiffen.

[10] Bernstein, S.; 2006. *Factorization of the nonlinear Schrödinger equation and applications.* Complex Variables and Elliptic Equations, **51**, No. 5–6, 429–452.

[11] Bernstein, S. and Gürlebeck, K.; 1999. *On a higher dimensional Miura transform.* Complex Variables **38**, 307–319.

[12] Bers, L.; 1950. *The expansion theorem for sigma-monogenic functions.* American Journal of Mathematics **72**, 705–712.

[13] Bers, L.; 1952. Theory of pseudo-analytic functions. New York University.

[14] Bers, L.; 1956. *An outline of the theory of pseudoanalytic functions.* Bull. Amer. Math. Soc. **62**, 291–331.

[15] Bers, L.; 1956. *Formal powers and power series.* Communications on Pure and Applied Mathematics **9**, 693–711.

[16] Bjorken, J. and Drell, S.; 1998. Relativistic quantum mechanics. The McGraw-Hill Companies, Inc.

[17] Bliev, N.; 1997. Generalized analytic functions in fractional spaces. Addison Wesley Longman Ltd.

[18] Bogdanov, Y. Mazanik, S. and Syroid, Y.; 1996. Course of differential equations. Minsk: Universitetskae (in Russian).

[19] Brackx, F., Delanghe, R. and Sommen, F.; 1982. Clifford Analysis. Pitman Publishing, Marshfield, MA.

[20] Bragg, L.R. and Dettman, J.W.; 1995. *Function theories for the Yukawa and Helmholtz equations.* Rocky Mountain Journal of Mathematics **25**, No. 3, 887–917.

[21] Castañeda, A. and Kravchenko, V.V.; 2005. *New applications of pseudoanalytic function theory to the Dirac equation.* J. of Physics A: Mathematical and General **38**, No. 42, 9207–9219.

[22] Castillo, R. and Kravchenko, V.V.; 2003. *General solution of the fermionic Casimir effect model.* Bull. de la Société des Sciences et des Lettres de Lódz, **53**, Série: Recherches sur les déformations, No. 41, 115–123.

[23] Chadan, Kh. and Kobayashi, R.; 2006. *New classes of potentials for which the radial Schrödinger equation can be solved at zero energy.* J. of Physics A: Mathematical and General **39**, No. 13, 3381–3396.

[24] Chadan, Kh. and Kobayashi, R.; 2006. *New classes of potentials for which the radial Schrödinger equation can be solved at zero energy:* II. J. of Physics A: Mathematical and General **39**, No. 44, 13691–13699.

[25] Chanane, B.; 1998. *Eigenvalues of Sturm-Liouville problems using Fliess series.* Applicable Analysis 69, 233–238.

[26] Chanane, B.; 2008. *Sturm-Liouville problems with parameter dependent potential and boundary conditions.* J. Comput. Appl. Math. 212 , no. 2, 282–290.

[27] Chemeris, V.S.; 1995. *Construction of Cauchy integrals for one class of non-analytic functions of complex variables.* J. Math. Sci. **75**, no. 4, 1857–1865.

[28] Child, M.S. and Chambers, A.V.; 1988. *Persistent accidental degeneracies for the Coffey-Evans potential.* J. Phys. Chem 92, 3122–3124.

[29] Code, W.J. and Browne, P.J.; 2005. *Sturm-Liouville problems with boundary conditions depending quadratically on the eigenparameter.* J. Math. Anal. Appl. 309, no. 2, 729–742.

[30] Colton, D.; 1980. Analytic theory of partial differential equations. Boston: Pitman Advanced Pub. Program.

[31] Coşkun, H. and Bayram, N.; 2005. *Asymptotics of eigenvalues for regular Sturm-Liouville problems with eigenvalue parameter in the boundary condition.* J. Math. Anal. Appl. 306, no. 2, 548–566.

[32] Courant, R. and Hilbert, D.; 1989. Methods of Mathematical Physics, v. 2. Wiley-Interscience.

[33] Daniljuk, I.I.; 1963. *A generalized Cauchy formula for axially symmetric fields.* (Russian) Sibirsk. Mat. Z. 4, 48–85.

[34] Davis ,H.; 1962. Introduction to nonlinear differential and integral equations. N.Y.: Dover Publications.

[35] Demidenko, Eu.; 2006. *Separable Laplace equation, magic Toeplitz matrix, and generalized Ohm's law.* Applied Mathematics and Computation **181**, 1313–1327.

[36] De Schepper, N. and Peña, D.; 2005. Factorization of the Schrödinger operator and the Riccati equation in the Clifford analysis setting. In: Liber Amicorum Richard Delanghe: een veelzijdig wiskundige, F. Brackx et al. (editors), Gent: Academia Press, 69–84.

[37] Duffin, R.J.; 1971. *Yukawan potential theory.* J. Math. Anal. Appl. **35**, 105–130.

[38] Duffin, R.J.; 1972. *Hilbert transforms in Yukawan potential theory.* Proc. Nat. Acad. Sci. **69**, 3677–3679.

[39] Dzhuraev, A.D.; 1987. Singular integral equation method. Moscow: Nauka (in Russian); Engl. transl. Longman Sci. Tech., Harlow and Wiley, N.Y., 1992.

[40] Gsponer, A. and Hurni, J.P.; 2001. *Comment on formulating and generalizing Dirac's, Proca's, and Maxwell's equations with biquaternions or Clifford numbers.* Foundations of Physics Letters **14** (1), 77–85.

[41] Gürlebeck, K. and Sprössig, W.; 1989. Quaternionic analysis and elliptic boundary value problems. Berlin: Akademie-Verlag.

[42] Gürlebeck, K. and Sprössig, W.; 1997. Quaternionic and Clifford Calculus for Physicists and Engineers. Chichester: John Wiley & Sons.

[43] Feng Qingzeng; 1997. *On force-free magnetic fields and Beltrami flows.* Applied Mathematics and Mechanics (English Edition) **18**, 997–1003.

[44] Fryant, A.; 1981. *Ultraspherical expansions and pseudo analytic functions.* Pacific Journal of Math. **94**, No. 1, 83–105.

[45] Fulton, Ch.T.; 1977. *Two-point boundary value problems with eigenvalue parameter contained in the boundary conditions.* Proc. Roy. Soc. Edinburgh Sect. A. 77, no. 3–4, 293–308.

[46] Gakhov, F.D.; 1966. Boundary value problems. Oxford: Pergamon Press.

[47] Garabedian, P.R.; 1956. *Calculation of axially symmetric cavities and jets.* Pac. J. Math. **6**, 611–684.

[48] Golubeva, O.V.; 1972. Course of mechanics of continuum media. Moscow: Visshaya Shkola (in Russian).

[49] Goman, O.G.; 1984. *Representation in terms of p-analytic functions of the general solution of equations of the theory of elasticity of a transversely isotropic body.* J. Appl. Math. Mech. **48**, 62–67.

[50] Gonzalez-Gascon, F. and Peralta-Salas, D.; 2001. *Ordered behaviour in force-free magnetic fields.* Physics Letters A **292**, 75–84.

[51] Gradshtein, I.S. and Rizhik, I.M.; 1963. Table of integrals, series, and products. Moscow: FizMatLit.

[52] Guo Chun Wen; 2003. Linear and Quasilinear Complex Equations of Hyperbolic and Mixed Type (Taylor & Francis London)

[53] Hille, E.; 1997. Ordinary Differential Equations in the Complex Domain. N.Y.: Dover Publications.

[54] Ismailov, A.Ja. and Tagieva, M.A.; 1970. *On the representation of generalized analytic functions by a series of pseudopolynomials.* Soviet Math. Dokl. **11**, No. 6, 1605–1608.

[55] Kaiser R., Neudert, M. and von Wahl, W.; 2000. *On the existence of force-free magnetic fields with small nonconstant α in exterior domains.* Communications in Mathematical Physics **211**, 111–136.

[56] Kamke, E.; 1976. Handbook of ordinary differential equations. Moscow: Nauka (Russian translation from the German original: 1959 Differentialgleichungen. Lösungsmethoden und Lösungen. Leipzig).

[57] Kantor, I.L. and Solodovnikov, A.S.; 1989. Hypercomplex numbers (Springer-Verlag New York)

[58] Kapshivyi, A.A. and Yazkulyev, M.; 1993. *Solution of boundary-value problems of p-analytical functions with the characteristic $p = x/(x^2 + y^2)$ on a halfplane with cuts.* J. Sov. Math. **66**, No.4, 2369–2376.

[59] Kapshivyi, O.O. and Klen, I.V.; 1995. *Solving boundary-value problems of x^k-analytic functions for a semicircle.* J. Math. Sci. **75**, No. 4, 1785–1791.

[60] Khmelnytskaya, K.V. and Kravchenko, V.V.; 2008. *On a complex differential Riccati equation.* Journal of Physics A, v. 41, No. 8, 085205.

[61] Khmelnytskaya, K.V., Kravchenko, V.V. and Oviedo, H.; *On the solution of the static Maxwell system in axially symmetric inhomogeneous media.* To appear in Mathematical Methods in the Applied Sciences.

[62] Khmelnytskaya, K.V. and Rosu, H.C.; 2009. *An amplitude-phase (Ermakov–Lewis) approach for the Jackiw–Pi model of bilayer graphene.* J. of Phys. A **42**, No. 4, 042004.

[63] Kravchenko, V.G. and Kravchenko, V.V.; 2003. *Quaternionic factorization of the Schrödinger operator and its applications to some first order systems of mathematical physics.* Journal of Physics A **36**, 11285–97.

[64] Kravchenko, V.G., Kravchenko, V.V. and Williams, B.D.; 2001. A quaternionic generalization of the Riccati differential equation. Dordrecht: Kluwer Acad. Publ., "Clifford Analysis and Its Applications", Ed. by F. Brackx et al., 143–54.

[65] Kravchenko, V.V.; 1995. *On a biquaternionic bag model.* Zeitschrift für Analysis und ihre Anwendungen **14** (1), 3–14.

[66] Kravchenko, V.V.; 2003. Applied quaternionic analysis. Lemgo: Heldermann Verlag.

[67] Kravchenko, V.V.; 2003. *On Beltrami fields with nonconstant proportionality factor.* J. of Phys. A **36**, 1515–1522.

[68] Kravchenko, V.V.; 2005. *On the reduction of the multidimensional stationary Schrödinger equation to a first order equation and its relation to the pseudoanalytic function theory.* J. of Phys. A **38**, No. 4, 851–868.

[69] Kravchenko, V.V.; 2005. *On a relation of pseudoanalytic function theory to the two-dimensional stationary Schrödinger equation and Taylor series in formal powers for its solutions.* J. of Phys. A , **38**, No. 18, 3947–3964.

[70] Kravchenko, V.V.; 2005. *On the relationship between p-analytic functions and the Schrödinger equation.* Zeitschrift für Analysis und ihre Anwendungen, **24**, No. 3, 487–496.

[71] Kravchenko, V.V.; 2006. *On a factorization of second order elliptic operators and applications.* Journal of Physics A: Mathematical and General, **39**, No. 40, 12407–12425.

[72] Kravchenko, V.V.; 2008. *On a transplant operator and explicit construction of Cauchy-type integral representations for p-analytic functions.* Journal of Mathematical Analysis and Applications, v. 339, issue 2, 1103–1111.

[73] Kravchenko, V.V.; 2008. Recent developments in applied pseudoanalytic function theory. In "Some topics on value distribution and differentiability in complex and p-adic analysis", eds. A. Escassut, W. Tutschke and C.C. Yang, Science Press 293–328.

[74] Kravchenko, V.V.; 2008. *A representation for solutions of the Sturm-Liouville equation.* Complex Variables and Elliptic Equations, v. 53, issue 8, 775–789.

[75] Kravchenko, V.V. and Oviedo, H.; 2003. *On a quaternionic reformulation of Maxwell's equations for chiral media and its applications.* Zeitschrift für Analysis und ihre Anwendungen **22**, No. 3, 569–589.

[76] Kravchenko, V.V. and Oviedo, H.; 2007. *On explicitly solvable Vekua equations and explicit solution of the stationary Schrödinger equation and of the*

equation div$(\sigma\nabla u) = 0$. Complex Variables and Elliptic Equations **52**, No. 5, 353–366.

[77] Kravchenko, V.V. and Oviedo, H.; 2008. *On Beltrami fields with nonconstant proportionality factor on the plane.* Reports on Mathematical Physics, v. 61, No. 1, 29–38.

[78] Kravchenko, V.V. and Porter, R.M.; 2008. *Spectral parameter power series for Sturm-Liouville problems.* arXiv:0811.4488, to appear in Mathematical Methods in the Applied Sciences.

[79] Kravchenko, V.V. and Ramirez, M.; 2003. *On a quaternionic reformulation of the Dirac equation and its relationship with Maxwell's system.* Bulletin de la Société des Sciences et des Lettres de Lódz **53**, Série: Recherches sur les déformations, No. 41, 101–114.

[80] Kravchenko, V.V., Rochon, D. and Tremblay, S.; 2008. *On the Klein-Gordon equation and hyperbolic pseudoanalytic function theory.* J. Phys. A: Math. Theor. 41, issue 6, 065205.

[81] Kravchenko, V.V. and Shapiro, M.V.; 1996. Integral representations for spatial models of mathematical physics. Harlow: Addison Wesley Longman Ltd., Pitman Res. Notes in Math. Series, v. 351.

[82] Lakhtakia, A.; 1994. Beltrami fields in chiral media. Singapore: World Scientific.

[83] Lanczos, C.; 1929. *The tensor analytical relationships of Dirac's equation.* Zeitschrift für Physik **57**, 447–473.

[84] Lavrentyev, M.A. and Shabat, B.V.; 1977. Hydrodynamics problems and their mathematical models (Nauka Moscow) (in Russian).

[85] Ledoux, V.; 2007. Study of Special Algorithms for solving Sturm-Liouville and Schrödinger Equations, thesis Universiteit Gent.

[86] Levitán, B.M. and Sargsjan, I.S.; 1991. Sturm-Liouville and Dirac operators. Dordrecht: Kluwer Acad. Publ.

[87] Madelung, E.; 1957. Die mathematischen Hilfsmittel des Physikers. Berlin: Springer-Verlag.

[88] Malonek, H.; 1998. Generalizing the (F, G)-derivative in the sense of Bers. Clifford Algebras and Their Application in Mathematical Physics (V. Dietrich et al. eds.), Kluwer Acad. Publ., 247–257.

[89] Marchenko, V.; 1988. Nonlinear equations and operator algebras. Dordrecht: Kluwer Academic Publishers.

[90] Matveev, V. and Salle, M.; 1991. Darboux transformations and solitons. N.Y. Springer.

[91] Menke, K.; 1974. *Zur Approximation pseudoanalytischer Funktionen durch Pseudopolynome.* Manuscripta Math. **11**, 111–125.

[92] Meziani, A.; 2008. *Representation of solutions of a singular Cauchy-Riemann equation in the plane.* Complex Variables and Elliptic Equations, v. 53, 1111–1130.

[93] Mielnik, B. and Reyes, M.; 1996. *The classical Schrödinger equation.* J. Phys. A: Math. Gen. **29**, No. 18, 6009–6025.

[94] Mielnik, B. and Rosas-Ortiz, O.; 2004. *Factorization: little or great algorithm?* J. Phys. A: Math. Gen. **37**, No. 43, 10007–10035.

[95] Motter, A.F. and Rosa, M.A.F.; 1998. *Hyperbolic calculus.* Adv. Appl. Clifford Algebras 8, no. 1, 109–128.

[96] Nachman, A.; 1988. *Reconstructions from boundary measurements.* Annals of Mathematics **128**, 531–576.

[97] Novikov, S.P. and Dynnikov, I.A.; 1997. *Discrete spectral symmetries of low-dimensional differential operators and difference operators on regular lattices and two-dimensional manifolds.* Russ. Math. Surv. **52**, No. 5, 1057–1116.

[98] Paine, J.W., De Hoog, F.R. and Anderssen, R.R.; 1981. *On the correction of finite difference eigenvalue approximations for Sturm-Liouville problems.* Computing 26, no. 2, 123–139.

[99] Pakhareva, N.A. and Belova, M.M.; 1972. *Inversion formulas for the fundamental integral representation of y^k-analytic functions in polar coordinates.* Ukrainian Mathematical Journal 24, No. 2, 228–231.

[100] Pakhareva, N.A. and Belova, M.M.; 1973. *On some boundary value problems in the class of y^k-analytic functions.* Ukrainian Mathematical Journal 25, No. 1, 37–45.

[101] Piven', V.F.; 1995. *Two-dimensional flow in porous strata with conductivity modeling by a harmonic function of the coordinates.* Fluid Dynam. 30, no. 3, 418–427.

[102] Piven', V.F.; 2006. *Singular integrals with Cauchy type kernels and their application to a two-dimensional problem on the evolution of a fluid interface in an inhomogeneous layer.* Differential Equations 42, No. 9, 1269–1282.

[103] Polozhy, G.N.; 1965. Generalization of the theory of analytic functions of complex variables: p-analytic and (p, q)-analytic functions and some applications. Kiev University Publishers (in Russian).

[104] Pöschel, J. and Trubowitz, E.; 1987. Inverse spectral theory. Boston: Academic Press.

[105] Price, G.B.; 1991. An introduction to multicomplex spaces and functions Marcel Dekker Inc. New York).

[106] Pryce, J.D.; 1993. Numerical solution of Sturm-Liouville problems. Clarendon Press.

[107] Reid, W.; 1972. Riccati differential equations. N.Y.: Academic Press.

[108] Rochon, D.; 2004. *A bicomplex Riemann zeta function.* Tokyo J. Math. 27, no. 2, 357–369.

[109] Rochon, D.; 2008. *On a relation of bicomplex pseudoanalytic function theory to the complexified stationary Schrödinger equation.* Complex Variables and Elliptic Equations, v. 53, No. 6, 501–521.

[110] Rochon, D. and Shapiro, M.; 2004. *On algebraic properties of bicomplex and hyperbolic numbers.* An. Univ. Oradea Fasc. Mat. 11, 71–110.

[111] Rochon, D. and Tremblay, S.; 2004. *Bicomplex quantum mechanics: I. The generalized Schrödinger equation.* Advances in Applied Clifford Algebras 14, No. 2, 231–248.

[112] Samsonov, B.F.; 1995. *On the equivalence of the integral and the differential exact solution generation methods for the one-dimensional Schrödinger equation.* J. Phys. A: Math. Gen. 28, No. 23, 6989–6998.

[113] Schwartz, Ch.; 2006. *Relativistic quaternionic wave equation.* Journal of Mathematical Physics, v. 47, no. 12, 122301.

[114] Sobczyk, G.; 1995. *The hyperbolic number plane.* Coll. Maths. Jour. 26, No. 4, 268–280.

[115] Thaller, B.; 1992. The Dirac equation. Berlin Heidelberg: Springer-Verlag.

[116] Tutschke, W.; 2003. Generalized analytic functions and their contributions to the development of mathematical analysis. Kluwer Acad. Publ., "Finite or Infinite Dimensional Complex Analysis and Applications" (Advances in Complex Analysis and Its Applications, 2), Ed. by Le Hung Son et al., 101–114.

[117] Tutschke, W. and Vasudeva, H.L.; 2005. An introduction to complex analysis: classical and modern approaches. Chapman & Hall/CRC.

[118] Uhlmann, G.; 1999. Developments in inverse problems since Calderón's foundational paper. In "Harmonic analysis and partial differential equations, Essays in Honor of Alberto P. Calderón". Chicago lectures in Mathematics, 295–345, edited by M. Christ, C.E. Kenig and C. Sadosky.

[119] Usmanov, Z.D.; 2006. *On characteristics of solutions to a model generalized Cauchy-Riemann system degenerating on a part of the boundary.* Complex Variables and Elliptic Equations, v. 51, 825–830.

[120] Vekua, I.N.; 1959. Generalized analytic functions. Moscow: Nauka (in Russian); English translation Oxford: Pergamon Press 1962.

[121] Walter, J.; 1973. *Regular eigenvalue problems with eigenvalue parameter in the boundary condition.* Math. Z. 133, 301–312.

[122] Watson, G.A.; 1922. A treatise on the Bessel functions. Cambridge: Cambridge University Press.

[123] Wendland, W.L.; 1979. Elliptic systems in the plane. London: Fearon-Pitman Inc.

[124] Weyl, H.; 1910. *Über gewöhnliche Differentialgleichungen mit Singularitäten und die zugehörigen Entwicklungen willkürlicher Funktionen.* (German) Math. Ann. 68, no. 2, 220–269.

[125] Yoshida, Z.; 1997. *Applications of Beltrami functions in plasma physics.* Non-linear Analysis, Theory & Applications **30**, 3617–3627.

[126] Zabarankin, M. and Krokhmal, P.; 2007. *Generalized Analytic Functions in 3D Stokes Flows.* The Quarterly Journal of Mechanics and Applied Mathematics, v. 60, no. 2, 99–123.

[127] Zabarankin, M. and Ulitko, A.F.; 2006. *Hilbert formulas for r-analytic functions in the domain exterior to spindle.* SIAM Journal of Applied Mathematics **66**, No. 4, 1270–1300.

[128] Zaghloul, H. and Barajas, O.; 1990. *Force-free magnetic fields.* American Journal of Physics **58**, (8), 783–788.

[129] Zettl, A.; 1997. Sturm-Liouville problems. In: Spectral theory and computational methods of Sturm-Liouville problems (Knoxville, TN, 1996), 1–104, Lecture Notes in Pure and Appl. Math., 191, Dekker, New York.

Index